The energy panaceas that were being advanced with confidence a decade ago are likely to be a lethal problem in themselves and no solution to any existing problem. Any nation that pursues the nuclear energy alternative not only increases the existing rate of fossil-fuel depletion, but further opens the path to nuclear war, nuclear blackmail and sabotage, the high risk of nuclear-power-plant accident, and finally the impossible task of finding a secure means for disposal of nuclear wastes. The nation that adopts the nuclear option helps to endanger the future of life on earth and almost guarantees the growing restriction of human freedom imposed by the need for increasing security measures. Furthermore, it is no answer to the energy problem, but may militate against finding long-term solutions.

— Raymond Dasmann
Toward a Dynamic Balance of Man and Nature
International Union for the Conservation of Nature and Natural Resources

Once a bright hope shared by all mankind, including myself, the rash proliferation of atomic-power plants has become one of the ugliest clouds overhanging America.

— David Lilienthal
First Chairman, U.S. Atomic Energy Commission

Uranium Mining in Northern Saskatchewan

Correspondence with the Premier

Bill Harding

International Year of the Child 1979

To our children
who will inherit
our madness
or our enlightenment

Typeset and printed in Canada by union labour.

Published by the Regina Group for a Non-Nuclear Society
2138 McIntyre Street
Regina, Saskatchewan

Typesetting and layout by Briarpatch Typesetting, Regina

Printed at Dumont Press Graphix, Kitchener, Ontario

Cover design and graphics by Denis Nokony

First printing April, 1979
Second printing July, 1979
Third printing October, 1979

Canadian Cataloguing on Publication Data
Harding, Bill, 1911-
Uranium mining in northern Saskatchewan

Bibliography: p.
ISBN 0-920850-00-6

1. Blakeney, Allan — Correspondence. 2. Harding, Bill, 1911- — Correspondence. 3. Uranium mines and mining — Saskatchewan. 4. Nuclear technology. 5. Atomic energy — Economic aspects.
I. Blakeney, Allan. II. Regina Group for a Non-Nuclear Society. III. Title.

TN490.U7H37 622'.3'493097124 C79-094276-3

Table of Contents

Foreword 7

An Open Letter to the Premier, September 20, 1978 11

Premier Blakeney's Reply, December 5, 1978 13

Mr. Harding's Response, March 15, 1979 17

- The Economics of Nuclear Energy 18
- The Demand for Nuclear Energy 23
- The Energy Needs of the Third World 30
- Nuclear Proliferation 44
- Radiation Standards and Public Health 53
- Nuclear Waste Storage 61
- Conclusion 67

Recommended Reading List 71

Interim Reply from Premier Blakeney's Deputy, October 12, 1978 73

Mr. Harding's Response, November 28, 1978 75

The Newspaper Advertisement referred to in Correspondence 80

Letters to "The Commonwealth" regarding the Ad 83

The splitting of the atom has changed everything save our modes of thinking and thus we drift toward unparalleled catastrophe.

— Albert Einstein

Foreword

The province of Saskatchewan contains one of the largest proven reserves of uranium in the world. The past few years have seen a high level of exploration activity and a number of unusual finds.

One such find at Cluff Lake in the Carswell Lake area, has a highly concentrated ore body. It is controlled by a foreign multinational, Amok, whose shareholders are three French private companies and a French state-owned corporation.

Controversy developed during 1976, particularly within the New Democratic Party, about the opening of a uranium mine at Cluff Lake. Subsequently, the NDP Government, on February 1, 1977, appointed a Board of Inquiry to recommend "whether the project should proceed, should not proceed, or proceed subject to specified conditions." Mr. Justice E.D. Bayda was appointed Chairman of the Inquiry.

The Board of Inquiry reported favorably on the project on May 31, 1978. Eight days later the Government accepted its major recommendations. The debate on this decision is continuing.

The exchange of letters on the uranium question which is presented in this book originated with an open letter from Bill Harding to Saskatchewan Premier Allan Blakeney when the recent provincial election was announced. Mr. Harding claimed he was effectively disenfranchised. The open letter also appeared in the Saskatchewan press.

8 Correspondence with the Premier

This open letter was acknowledged by the Premier's Office prior to the election. Understandably the Premier was not able to undertake a substantive reply until after the election. He did so on December 5.

The Premier's letter sets out the Government's position on six points related to uranium:

1. The economics of nuclear energy;
2. The demand for nuclear energy;
3. The energy needs of the Third World;
4. Proliferation of nuclear weaponry;
5. Radiation standards and public health; and
6. Nuclear waste storage.

In Mr. Harding's reply, he has dealt with each of these points in some detail. He speaks from wide experience. Raised in Swift Current, a graduate in agriculture, he has worked with both the Canadian and Saskatchewan Departments of Agriculture. He was Secretary of the Saskatchewan Royal Commission on Agriculture and Rural Life in the fifties. He spent twelve years as a development administrator with the United Nations Development Program (UNDP) in four countries: Liberia, Guyana, Somalia and the Philippines. Prior to his retirement in 1975 he was Director of Programme Policy for UNDP in United Nations Headquarters in New York.

Mr. Harding has kept abreast of the literature on nuclear energy since the late forties. While overseas he supervised a search for uranium in Somalia and a feasibility study for a nuclear power plant in the Philippines. He was **de facto** representative of the International Atomic Energy Agency (IAEA) for these and other activities. Mr. Harding retired and returned to his native Saskatchewan in 1975. He has been active in development education work since his return.

Mr. Harding has a close connection with the CCF. A supporter from the forties, he came to work with the first CCF government at the suggestion of the late Woodrow Lloyd, then Minister of Education. He was an active party member throughout this period, holding a number of elected constituency and provincial positions. He left the Government in 1962, assisted in the organization of Community Clinics at the time of the medicare crisis and then became Director of Education and Organization for the Party. The Premier and several members of the legislature are his former colleagues. This accounts for the frank manner in which they speak to one another in the correspondence.

The Regina Group for a Non-Nuclear Society (RGNNS) decided to publish this exchange of letters as a service to the public of Saskatchewan. The public has a right to know, not only the decision of the Government, but also considered criticisms of that decision. The Government has done little to inform the public about this issue. In fact, as the reader will see, the Government has gone to some lengths to keep the public uninformed.

RGNNS is happy to fill this need and to help increase public awareness of the deep problems surrounding the issue of uranium exploitation.

Finally, RGNNS wishes to thank the members of its Research and Literature Committee, particularly Ron Bocking, Don and Judy Gayton, Jim Harding, Helen MacFarlane, Brijesh Mathur and Maisie Shiell, for the assistance given to Mr. Harding in the preparation of his response. In addition, Mr. Harding wishes to thank Fraser and Jaki Russell, Gary Robins, Rici Liknaitzky, Helmut Kuhn, Pam Hanna, Laurie Beingessner and Debbie Abood for their generous co-operation at critical points in the production of this publication.

RGNNS will be most grateful for responses from the readers of this important dialogue.

Regina Group for a
Non-Nuclear Society
2138 McIntyre Street
Regina, Saskatchewan S4P 2R7

WELCOME TO BEAUTIFUL
NORTHERN SASKATCHEWAN
DANGER
RADIOACTIVE ZONE

An Open Letter to the Premier

2230 Smith Street
Regina, Saskatchewan
September 20, 1978

Dear Mr. Blakeney:

You have announced an election. By doing so you have very effectively disenfranchised me. Many many others face the same dilemma. Our personal dignity has been affronted.

Let me explain.

You know that I could never vote for the traditional free enterprise parties because they place profits and corporate control of the economy above the satisfaction of basic human needs and the freedom and liberty of people.

The CCF was known as a party of "Humanity First". It pioneered many socially just programs for Saskatchewan and, indeed, for all of Canada.

In making the incredible decision to facilitate and expand the exploitation of uranium in the interest of short-term economic gain for the province you have turned your back on all that the CCF and NDP have meant and done in the past.

Your narrow "golden egg" decision overlooks the fact that this uranium will go into a global nuclear cycle and will inevitably add to the stockpile of nuclear weapons. This will increase the repression of people around the world. It will increase the already imminent threat of nuclear war and thus the security of every Saskatchewan family. In fact, the balance between war and peace is now so delicate that the addition of Saskatchewan uranium to world supplies could well tip the scales of world history to the frightening reality of a nuclear holocaust.

Even more important is the effect of your decision about uranium development on the present and future health of Saskatchewan citizens. On this question the Bayda Board of Inquiry did you a great disservice. Its coverage of the health research available was inadequate and the analysis of the material was extremely superficial. People who are eminently qualified in the field of low level radiation state that present regulatory standards are arbitrary, that the incidence of cancer from low level radiation is very much higher than was previously thought **and that cellular damage suffered from low level radiation is transmittable to our offspring.**

So that, in addition to highly poisonous radioactive wastes (which will last for 3,000 generations) being added to our environment we will also be adding the seeds of genocide to our own bodies.

You did not know this before because you did not want to listen and because the "reputable" Board of Inquiry did not care to listen.

Please hear us now before it is too late. You yourself have said Saskatchewan doesn't need uranium. Be the statesman which you are pictured to be. Have the courage to admit an error and provide the leadership many think you capable of.

I can't vote Liberal or Conservative. Now I can't vote NDP because uranium development negates every worthwile advance the CCF/NDP ever made.

Please give me back my vote!

Yours for Humanity First,

Bill Harding

Bill Harding

Premier Blakeney's Reply

Legislative Building
Regina, Saskatchewan
December 5, 1978

Mr. Bill Harding,
2230 Smith Street,
Regina, Saskatchewan.

Dear Bill:

Wes Bolstad mentioned in his letter to you that I would be responding personally to your open letter of September 20th. I have now had time to give careful thought to your arguments.

In your letter you present two arguments against uranium mining: first, that it increases the risk of nuclear war and, second, that it constitutes a health hazard both from the effects of low-level radiation and from the long-range uncertainty of waste disposal.

Let me put those arguments aside for a moment and remind you that there are arguments in favour of uranium mining also, and not only of the "golden egg" variety. The generation of electricity by nuclear energy is proving, in many parts of the world, to be the most economical source of energy. I recognize that alternative, safer, sources of energy may be generally available in a few decades. Every projection I have seen however, indicates that nuclear energy will be essential for at least some period of time, even with stringent conservation measures and greatly expanded research into alternative energy sources.

I am also familiar with the argument that third world countries should not bind themselves to the sophisticated technology of nuclear power plants. My response to that is that for many third world countries, the only energy available for industrial development is nuclear energy, and that to refuse them nuclear energy is to condemn them to a future of subsistence living. I do not believe that we, in energy-rich Canada, have the right to make that choice for other people.

It is widely acknowledged that if countries like Japan do not use nuclear power, they will generate the power they believe they need from coal and oil. The effect on less developed countries which will have to compete with Japan for scarcer fuel sources will be highly adverse. Since it is in the highest degree unlikely that the Government of Saskatchewan could convince the well-off nations to refrain from using the power sources needed by less developed countries, we can withhold our uranium only if we are prepared to condone these adverse effects on the less developed countries. The increase in oil prices has already caused great hardship. The continued increase in prices of oil and coal which would result if Saskatchewan uranium was withheld (unless it is replaced by other uranium) would cause even greater hardship and suffering. I think that course of action is not justifiable on moral grounds.

I consider my position to be a moral position, as deeply felt as yours. It is not, as you suggest, based simply on the prospect of all those lovely uranium dollars.

When you argue that our uranium will "inevitably" add to the stockpile of nuclear weapons, you take a very dark view of human nature. Medical research greatly increases the possibility of biological warfare, in some ways an even more frightening prospect than nuclear warfare. Space research has obviously been corrupted for military purposes, some highly lethal. Yet there is no cry to stop those forms of endeavour. We continue to hope that sane minds will prevail, and that the great military machines can be stopped by our collective will. They will not be stopped by a shortage of uranium. There are weapons enough available now to destroy the world totally. The balance you speak of has long since been tipped.

We should not prevent the beneficial peaceful uses of uranium simply because uranium is a component of nuclear weapons. The prevention of nuclear war lies in the political and moral sphere, and will not be affected one way or another, I believe, by a ban on uranium mining.

Your second point is more difficult to handle. I am not a scientist, but I have been around long enough to see scientific opinion change dramatically with regard to X-rays and DDT and a variety of other processes and substances once thought to be beneficial. It is entirely possible that further research will show that our regulatory standards are inadequate. If that happens, we will change our standards.

At the moment, we have the toughest standards in North America, and those standards are based on the best information available to us. It is always a little chilling to talk about "acceptable" levels of risk, but in any industrial enterprise, particularly mining, there is bound to be some risk. We believe that, given our present standards, uranium mining in Saskatchewan is less risky than coal mining or asbestos mining, to name just two obvious examples. We do not believe the risk is sufficient to stop the mining.

On the issue of radioactive waste disposal we have simply had to make a leap of faith and assume that a satisfactory means of disposal will shortly be found. The waste created by nuclear power generation is only a tiny fraction of the total problem, as you know, so it did not seem to us very useful to stop the generation of nuclear power when that gesture would not appreciably diminish the problem of nuclear waste.

The issues you raise are tough issues, and we do not treat them lightly. They are moral issues, as you rightly point out, and we have taken a moral position on them. I know your conscience compels you to disagree. I simply want you to understand that for many of us the government's position on uranium mining and the old slogan of "Humanity First" are not incompatible.

Thank you for speaking out so frankly. I have tried to reply in kind.

Yours sincerely,

Allan Blakeney, Premier.

There is no secret and there is no defence; there is no possibility of control except through the aroused understanding and insistence of the people of the world.

— Albert Einstein

Mr. Harding's Response

2230 Smith Street
Regina, Saskatchewan
March 15, 1979

Premier A. E. Blakeney
Legislative Building
Regina, Saskatchewan

Dear Al:

I appreciate your letter of December 5 in reply to mine of September 20. I also wish to acknowledge the interim reply from your deputy, Wes Bolstad, on October 12.

Your letter deals with six points:

1. The economics of nuclear energy;
2. The demand for nuclear energy;
3. The energy needs of the third world;
4. Proliferation of nuclear weaponry;
5. Radiation standards and public health; and
6. Nuclear waste storage.

These issues are so basic that they must be explored in depth. Otherwise our differences will be seen as rhetorical and will not contribute to public understanding. This accounts for the length of this letter.

My letter of September 20 was an open letter. I intend to publish this answer as a further open letter, along with your letter of 5 December, to which it is addressed. My purpose is to seek greater public understanding of the important question we are discussing.

I will now deal with the six points you raise:

1. The Economics of Nuclear Energy

> **You state:** "The generation of electricity by nuclear energy is proving, in many parts of the world, to be the most economical source of energy." (paragraph 3)

Your statement fails to make an important distinction that underlies sound energy planning. Electrical energy is actually the most expensive kind of energy. The losses in the conversion of the primary energy sources used to produce and distribute electricity waste around fifty percent of the primary fuel's energy potential. This is an extremely inefficient way to produce energy.

Nuclear power can presently only be used for the production of **electrical** energy. In this way it is more restricted than fossil fuels, solar power, etc. Electrical energy currently supplies only 14 percent of Canada's total energy consumption. Further, electricity is now used for purposes for which it is not the most efficient energy source, such as home heating. Sound energy planning will work to decrease our dependence on electricity as an energy source. Its share of total energy consumption may be reduced over time to five percent.

I am assuming therefore that you intended to say that "nuclear energy is proving... to be the most economical source of **electrical** energy". There are, however, several reasons for questioning this statement.

The most significant feature of projections of nuclear power capital costs is their serious underestimation of these costs. In 1967 the U.S. Atomic Energy Commission (AEC) estimated capital costs for nuclear power plants at $134 per kilowatt (KW). By 1976, 1000 megawatt (MW) plants slated for delivery in 1985 were expected to cost $1200/KW. A study from Harvard Business School and Massachusetts Institute of Technology (MIT) has shown that the average annual dollar increase in nuclear capital costs has been more than twice the average increase for coal plants during the period 1969 to 1975.

Moreover, they found that the capital costs of nuclear power are increasing at a rate of $19/KW per year more than the costs of coal. They conclude: "The capital costs of light-water reactors show no signs of stabilizing and, indeed, are apparently still climbing at alarming rates." (Irvin C. Bupp et al., "The Economics of Nuclear Power", **Technology Review**, February 1975 — from the book **Nuclear Power: The Unviable Option**, John J. Berger, Ramparts Press)

Noting the nearly tenfold increase in nuclear capital costs that has occurred in the past decade, Daniel Ford, executive director of the Union of Concerned Scientists, estimates that capital costs

---

A U.S. study prepared in December 1975 by the prestigious Argonne National Laboratory, points out that CANDU capital costs are much higher than those for a comparable U.S. reactor, amounting to a penalty of about $175 per kilowatt installed.

---

for nuclear power plants starting up in the early 1980s may well be $1500 to $2000/KW. Ford expects nuclear cost escalation to continue because of current and anticipated safety-related plant design and construction regulations. (**Nuclear Power: the Unviable Option**, cited above)

These spiralling capital costs undoubtedly account for the dramatic decline in the nuclear power construction industry. The following figures come from the reputable Kidder Peabody "Status Report on Electric Utility Generating Equipment: 1977 Orders", February 17, 1978.

The peak period for orders of nuclear plants by the utilities of the U.S. was prior to 1974. During 1973 and 1974 orders were placed for a total of nearly 60,000 MW of nuclear installations. However, the picture began to radically reverse during 1974. While there were 24 new orders placed in 1974 there were also cancellations of 20 orders that had been placed in previous years. In 1975 there were five new orders and 16 cancellations, plus 47 deferrals of more than one year. In 1976 there were only two new orders along with 11 cancellations and 51 postponements of over one year. In 1977 there were four new orders, less seven cancellations or indefinite deferrals and another 19 postponed for a year or more.

The four plants ordered in 1977 could hardly be considered "new". One was the exercise of an option placed in 1973 by the Arizona Public Service. Two were replacement orders by New York State Electric and Gas for units first ordered in 1971 and cancelled in 1975.

There were no new orders in 1978. Thus there have now been four years during which there has been negative growth in the industry. An update on the conditions of the industry is given in a Special Report published in **Business Week**, December 25, 1978. It says: "Not soon, but within ten years, the U.S. nuclear industry is apt to contract dramatically, and it may collapse altogether."

This does not mean U.S. utilities have not been constructing new plants. While orders for nuclear plants have been declining, orders for coal plants have increased (from 36 percent of total orders in 1973 to 63 percent in 1977). This is concrete evidence of where the economic benefits lie, for the utilities will invest in the cheapest source of electricity.

The advantage claimed for nuclear power is in the area of lower fuel costs, hence lower operating costs. This is a very hypothetical argument because, with the oil companies gaining ever greater control over both coal and uranium supplies, and through price fixing by the uranium cartel, we can anticipate that the price of the various supplies will be manipulated on the world market in such a way as to maximize profits by trying to keep all modes in operation. In any case, the utilities in the U.S. have clearly kept fuel and operating costs in mind in their recent dramatic shift to coal.

A critical component in estimating the overall cost of electricity from various primary energy sources is the estimate one places on the capacity of the various systems (the percentage of the rated capacity at which the system will actually produce electrical power). In regard to nuclear power, the industry maintains that nuclear power plants will operate at 75 to 80 percent of capacity. In the U.S. the actual experience is that nuclear power plants have operated at only 57.6 percent of capacity. (**Not Man Apart**, Friends of the Earth, April, 1978)

Another inadequacy of comparative cost studies of electrical power is that they usually compare only the traditional types of production. At the Conference on Canadian Nuclear Policy sponsored by Carleton University School of Public Administration in November 1978, a paper on "The Economics of Canadian Nuclear Policy" was presented by Professor S. Banerjee of

McMaster University. It was simply a comparison between nuclear and coal. There was no attempt to compare other traditional forms (e.g. oil, gas, hydro) or renewable alternatives. There was no attempt to compare the economics of various mixes of power sources. This particular paper concluded in favor of nuclear; however, this type of analysis is so narrow in its conception that its conclusions are virtually pre-ordained and therefore without validity or value.

In view of the above I would be most interested in learning where you obtain the evidence that "in many parts of the world" nuclear power is "the most economical source" of electrical energy.

There is another very important aspect to this question of the costs of nuclear power. Every analysis that I have seen which suggests a superiority for electricity from nuclear power overlooks many of its real costs. For example research and development (R & D) are seldom included as a cost of nuclear power. In the U.S. the public purse has provided $17 billion for R & D to the nuclear industry (not including unknown billions for military R & D which have had major power plant spinoffs). These costs are not included in the traditional cost/benefit analysis of nuclear power in either the U.S. or Canada. (The R & D costs in Canada are in the hundreds of millions of dollars.)

Another cost borne by the public and not included in the traditional cost/benefit analysis is the cost of government agencies whose sole purpose is the development, promotion, supervision and regulation of nuclear technology. In Canada, the Atomic Energy of Canada Limited (AECL) and the Atomic Energy Control Board (AECB) cost the taxpayers over $100 million each year.

Another cost which is subsidized by the public is the nuclear industry's insurance costs. Insurance companies will not provide unlimited coverage to nuclear plants because of the untold costs which would result from a major reactor accident. Governments now subsidize these costs.

Then there are subsidies related to the sale of nuclear reactors to other countries. These are in the form of payments made both above and below the table. The most recent example of the former is the approved Canadian sale of four nuclear reactors to Romania. To facilitate this sale, a one billion dollar loan to Romania, the largest loan of its kind, was recently approved by Canada's Export Development Corporation. The interest rate and repayment terms will likely be better than the going market rates: thus, another subsidy.

We do know about some of the payments made below the table: the most notorious being the multi-million dollar payment by AECL in connection with the sale of the CANDU to Argentina.

You may be reluctant to include these legal and illegal costs for the export of nuclear technology as legitimate charges against nuclear power **per se**. However, they are very important subsidies to the nuclear manufacturing industry in Canada. This industry requires a minimum of two new orders per year to stay alive: by 1980 it is estimated that the industry will need four new orders a year to be viable. Therefore, these subsidies must be considered as a direct cost of the nuclear industry.

I have said nothing about the costs of research and management of waste storage, decommissioning nuclear plants or public health. As there is no known safe method of waste storage, this cost can not be determined: if a large-scale plan is ever developed it may be only at a prohibitive cost. One small experimental plant has been decommissioned: no commercial plants have been decommissioned. The experimental plant which was done cost as much to decommission as it did to build! The most informed guess of the cost of decommissioning commercial plants is in excess of $100 million. I will say something about public health costs later.

I conclude that the economic analyses of nuclear power which show it in an advantageous position never include the costs of public support to the industry.

Did you ever stop to consider the subsidy which Canadian taxpayers have paid to build the nuclear power industry in Ontario? This might be a good topic for a future meeting of the First Ministers.

The long-held dream that nuclear power would give the United States and the world an endless stream of low-cost electric power has faded, according to a growing number of economists, technical experts and utility officials.

— David Burnham, "Hope for Cheap Power From Atom Is Fading", *New York Times*, November 16, 1975

---

2. The Demand for Nuclear Energy

> **You state:** "Every projection I have seen... indicates that nuclear energy will be essential for at least some period of time, even with stringent conservation measures and greatly expanded research into alternative energy sources." (paragraph 3)

I am simply astounded by this statement. To say "every projection" you have seen says that nuclear energy will be essential for at least some period of time is simply not correct. I personally sent you a copy of Amory Lovins' book, **Soft Energy Paths**, which makes quite a different projection, in March 1978. You acknowledged its receipt.

Amory Lovins is an internationally recognized consultant physicist. His current or recent clients include the Organization for Economic Co-operation and Development (OECD), several United Nations agencies, the International Federation of Institutes for Advanced Study, the MIT Workshop on Alternative Energy Strategies, the Science Council of Canada, Petro-Canada, the U.S. Energy Research and Development Administration (ERDA), the U.S. Office of Technology Assessment and other organizations in several countries. I think it is fair to say that he is one of the most renowned experts on the question of comparative costs of various energy scenarios.

Mr. Lovins was an expert witness at the Cluff Lake (Bayda) Inquiry. The Bayda report completely misrepresented his testimony as we shall document in our critique of the Bayda report now in preparation. Furthermore Mr. Lovins' time was available to the Saskatchewan Cabinet while he was in Regina as a witness and you and your colleagues neglected to take advantage of this. I can only conclude that there is presently no commitment to become acquainted with ideas and facts about alternative renewable energy policies in your rush to sell uranium.

The projections you refer to come from the energy corporations and from bureaucrats whose major concern is to maintain the present trends in the energy sector of western industrialized societies. The energy companies, after all, are the source of most of the data on which government supply projections are made. They have a history of telling governments what they want governments to know so that projections will favour the interests of the energy companies. (Often it is also in government's narrow interests to accept these projections.) This is the only explanation for the wide fluctuations in the projections made by the National Energy Board (NEB) over the past few years. The most recent example is that in 1974 the NEB said supplies of oil would run out in 1985; in 1978 it said 1995. The most notable result of these variable projections has been a very dramatic rise in the profits of the oil companies since the Organization of Petroleum Exporting Countries (OPEC) price decision in 1973 and the crisis mentality which has been fostered following that action.

It is important to recall that there have been many previous times when oil and energy crises have been proclaimed. In 1908 the U.S. Geological Survey estimated maximum future supply of oil at 23 billion barrels. The U.S. Bureau of Mines reduced this to six billion barrels in 1914. The Director of the U.S. Geological Survey announced that peak domestic production had almost been reached in 1920. In 1939 the U.S. Department of the Interior forecast that U.S. oil supplies would last only 13 years.

Energy Supplies

The facts are that there is no impending doom about traditional energy supplies. Let me cite a few examples.

Saudi Arabia has enough oil to last 150 years at present rates of production and export. It is presently using only 15 of its 37 known oil fields. Every year it has discovered more oil that it has produced.

The new discoveries in southern Mexico provide reserves of 60 billion barrels: twice total U.S. reserves.

The Alberta Resources Conservation Board has estimated reserves of 600 billion barrels (greater than Saudi Arabia's reserves) in the Athabaska oil sands. Some studies place it at one trillion barrels.

The press has been full of reports for the past year of dozens of new finds, new promises and upward revisions of reserves of oil, gas and coal — in Canada and elsewhere.

The main reason why supply estimates have gone up so much recently is because of the constant rise in price. Prices will continue to rise (witness the recent OPEC decision to increase the price another 14.5 percent in 1979) and this will stimulate the production of new technology and new finds. There is no question but that the end of these finite supplies will arrive some day in the future but **there is no reliable evidence that that time is near at hand or is so short that we must turn to nuclear energy**.

Probably the most prestigious conference on this question was held in Laxenburg, Austria in July 1976. It was sponsored jointly by the United Nations Institute for Training and Research (UNITAR) and the International Institute for Applied Systems Analysis (IIASA). The subject was "The Future Supply of Nature-Made Oil and Gas". Ninety scientists, engineers, geologists, economists and other specialists attended. The technical reports presented to the conference were published by Pergamon Press ($60). The key finding of this conference was that "while there was general agreement that oil and gas from (conventional) sources would last at least until about the time period 2020-30, it was also expected that some resources not now being exploited would by then be contributing increasingly to global supplies". The Alberta, Saskatchewan and Venezuela tar sands now being developed were in the non-conventional category. Including these and others, the conference found there was enough nature-made oil and gas known to exist at present to last the world for one hundred years.

This report has received virtually no public exposure. Bayda himself paid no attention to it although I brought it to his attention when I was a witness. Instead, the public has been exposed to energy company advertising telling them of the impending energy disaster facing mankind, the nuclear industry telling how nuclear power is essential to bridge a "gap", and bureaucrats going along uncritically with such tactics by the energy companies.

A great mythology has been created on this subject and what disturbs me the most is that politicians from all parties have accepted the myth and help to propagate it. Your second statement, quoted above, must make the oil and nuclear industry, which is now highly integrated, very happy indeed!

Energy Demand

The other side of the question is that while supply responds to price, so does demand. Many recent projections of demand have been based on pre-OPEC experience. Growth in electricity demand, however, has declined from the experience of that time. For example, in New York State in 1973 the utility companies projected a peak demand of 27,000 MW for 1977; in fact the demand reached only 21,000 MW.

The Ontario Royal Commission on Electric Power Planning (the Porter Commission) in its interim report "A Race Against Time" presents the most recent evidence on this point. After pointing out that Ontario Hydro's 1977 forecast of annual growth rate in demand was 6.2 percent (page 21) it concludes:

> On the basis of present incomplete knowledge, our "interim conclusion" is that the most probable electrical growth rate to 2000 is in the 4.0 percent growth rate.... The 2 percent growth rate is also credible. (page 23)

Conservation and Alternative Energy Potentials

The potential of savings from a truly national effort in conservation and the development of renewable alternatives has been ably outlined by Lovins and many others knowledgeable in those fields. Let me merely refer to two pieces of information on the potential of renewable energy which you may not have seen.

1. In April 1978 the U.S. President's Council on Environmental Quality estimated that, with a strong national effort, solar technology could supply a quarter of all U.S. energy needs by the year 2000 and "significantly more than half" by 2020.

2. Solar energy researchers at the Unversity of Waterloo say that widespread use of solar heating in Canada has now become economically feasible.

You may have seen the CBC Quarterly Report with Barbara Frum on December 17. This had encouraging new information on the immediate prospects for solar power.

Anything that slows down the development of solar energy — the one cheap, limitless source of energy that cannot be shut down by war or embargo — is undermining the national security.

— *The Role of Small Business in Solar Energy Research, Development, and Demonstration,* Interim Report of the Select Committee on Small Business, U.S. Senate, October 9, 1975.

The CBC program also presented a good deal of information on the potentials of conservation. Everyone is conservation conscious these days and governments have begun to promote and modestly support energy conservation measures. However, the efforts in total are largely cosmetic: they do not constitute a substantial and comprehensive program. If there was broad public understanding and enlightened political leadership on this question we could extend the life of non-renewable fuels **and** reduce the investment in new power plants very substantially.

First priority in any long-range energy policy must go to conservation, to reducing the demand side of the energy equation by eliminating wasteful practices and improving the efficiency with which energy is produced, delivered and employed. Study after study has shown conservation of existing energy resources to be the most effective and readily available "new source" of supply to meet urgent needs.

— The New York Times

Energy Policy and Employment

An important aspect of this question is the effect of solar power development and conservation activities on employment. Labour has tended to support large scale energy projects because of the growing shortage of jobs. However industry and government have never made the relative employment capacity of different energy policies a fundamental consideration. As a result, labour has never had an alternative choice.

Capital investment per job created in the nuclear industry runs several times higher than capital investment in conservation or in the development of renewable energy resources. A study of the Long Island, New York region, carried out by the U.S. Council on Economic Priorities, shows that an investment in solar energy and conservation would provide nearly three times as many jobs and save or produce as much energy as the investment of an equal amount in a nuclear power plant (not including the hidden costs mentioned earlier). Edward Carlough, president of the Sheetmetal Workers International Union, has said: "Even figured conservatively, energy-saving modification work and an expanded

Structural inflation and structural unemployment are related to the excessive capital intensity, energy intensity and materials intensity of these mature industrial economies. They are not labor intensive enough. They use natural resources too profligately, and they waste human resources and leave them unemployed.

There are all kinds of different price tags for jobs. For instance, jobs in nuclear power are incredibly expensive to create with the taxpayer's money, whereas jobs in the construction industry, to retrofit buildings to conserve energy, where you end up with the same number of BTU's as if you had built a power plant, give you a much bigger bang for your buck.

— Hazel Henderson, Director, Environmentalists for Full Employment;
Co-director, Princeton Center for Alternative Futures.

use of solar energy could put all unemployed sheetmetal workers back to work.'' **The employment potential is probably one of the most important considerations in the entire question of energy policy.**

Summary

Let me summarize my response to your second point:

1. There is no imminent shortage of traditional, non-renewable supplies of energy. The reverse is probably closer to the truth. Promising new finds are occurring, and new technology is emerging with greater frequency as prices increase.

2. The rate of increase in demand for electricity is declining.

3. Consumption of energy can be substantially reduced through conservation measures.

4. The technology is already available and quickly being put into use for a variety of solar, wind and other **renewable** forms of energy.

In view of this I would ask you, who are the people appealing to people's insecurities and fears in this situation? Are they the concerned citizens who feel uneasy about their environment, their health, the welfare and the peace of their world? Or are they the industrialists and their public relations spokesmen who tell people they will soon freeze in the dark unless industry builds nuclear plants?

The only ones who benefit from the mythology of scarcity, and support of this from government, are the integrated multi-national energy corporations and those who identify with their interests.

Surely the time has come for serious public discussion of these issues. The Bayda Inquiry made virtually no attempt to address these fundamental questions, although considerable evidence was provided to the Inquiry on the subject. In fact the Inquiry started with certain assumptions which effectively ruled out serious consideration of the question of energy supplies. For example, in the preface of its report it stresses the importance of nuclear "fission for producing **much needed** (my emphasis) electricity and energy" (Paragraph P5). In Chapter I the report talks about "fast disappearing fossil fuels" and "ever increasing new demands for energy", and determines that "renewable energy, particularly solar energy (is) now only in the rudimentary stages of development" (paragraph 1.5). This dismissal is hardly the way to discuss questions of such critical importance. **These biases and omissions of the Bayda Inquiry bring into serious question its conclusions and recommendations and the government's decision on uranium exploitation.**

Over the next few decades we could reduce the energy required to make things by a third, reduce the energy required to run homes and buildings by 40 percent, and reduce transportation needs by 20 to 40 percent. Aside from the changes in cars and in added recycling, we'll hardly notice the change in energy use except at the bank; for in addition to saving millions of barrels of oil each day we'll save billions of dollars a year.

— Lee Schipper, Energy Resources Group
University of California, Berkley.

---

3. The Energy Needs of the Third World

You state: "to refuse (third-world countries) nuclear energy is to condemn them to a future of subsistence living". (paragraph 4)
You talk about the difficulty of third-world countries competing with developed countries (especially Japan) for energy supplies. (paragraph 5)
You emphasize what you call the moral nature of your position in support of nuclear energy. (paragraph 6)

This point is primarily based upon the myth of energy supply shortage which I have already addressed. It is clear from what I have said that there is no imminent shortage in spite of what the energy companies have told us. Therefore I can place no credibility whatever in your protestations on behalf of the third world.

If the problem is not one of supply and the competition for short supplies, then it clearly must be one related to the distribution of supplies. To deal adequately with this question I must discuss the development process in the developing countries and relate this to the third world's real energy needs and to the question of the distribution of energy supplies.

The Conditions of Underdevelopment

The underlying cause of the third world's situation is the historical development of third world economies as colonies by European, and more recently, North American powers. The colonizing countries of Europe contended with each other to capture monopolistic control over the economies of the societies of Asia, Africa and South America. Having done this, they systematically proceeded to extract the resources of the colonized region using the cheap energy of forced labour. The character and intent of this relationship was fully exposed when the people of the colonized regions were taken captive as slaves and exported as commodities to serve wealthy masters. The effect of this relationship on the colonized countries was to destroy their political structures, to undermine their cultures, to rob them of their human resources and their dignity, and to make their once self-reliant developed (relative to Europe) societies underdeveloped.

Conversely, the economies of Europe and North America, the now industrialized western countries, were built to a large extent on the resources of the colonized countries, on their products, their raw materials and their labour, always undervalued by the force of colonial law and colonial arms. **This is the origin of the development/underdevelopment dilemma which plagues the world to this day.**

The wave of independence which followed World War II carried the promise of ending this state of inequity. It didn't take long, however, before the third world realized that political independence had little to do with real economic independence. Third-world economies were so integrated into the market economy controlled by the western industrialized countries and into international institutions, particularly the International Monetary Fund (IMF), that they found it virtually impossible to extricate themselves and improve their situation. In fact their situation has worsened over the years because of the tremendous growth and strength of the multinational corporations which, with the support of their home governments, have consolidated and strengthened this global economic system to the greater benefit of the industrialized economies.

From the fifties until the early seventies, many serious efforts were made to improve the situation of the underdeveloped countries. At the heart of these efforts was the goal of industrialization. The widespread assumption was that the poor countries could better their lot in the same manner as the developed countries had done. In this scenario, supported by most western economists, the developing countries, with appropriate help could, through industrialization (meaning capitalism in this context), achieve a growth rate which would reach a "take-off" point and ensure continuous growth until they reached acceptable equality with the modernized economies. The benefits in the short run would go to the propertied class in the developing nations but this was considered to be a good thing as this class would plough its profits back into enterprises which would ensure that growth continued. In any case it was believed that the benefits would increasingly "trickle down" to the poor masses.

The Saskatchewan Government's approach to northern development, in essence, is based upon the same "trickle down" theory. Northern Saskatchewan has many of the characteristics of the third world. Your government is opening it up rapidly for resource extraction. Granted, you supplement this program with modest efforts towards self-government and social assistance but the underlying consideration is resource extraction, presumably justified by the hope that native people will gain employment opportunities, that the benefits will "trickle down" to them. According to this line of thinking the "native problem" will disappear when all native people have been assimilated into this section of Western industrialized society. This is nothing more than modern-day imperialism.

At the international level, a whole range of programs were developed to encourage the desired industrialization. These included manpower training, import substitution, export promotion, commodity assistance, feasibility studies, infrastructure development, increased capital flows, etc., etc.

I could comment from first hand experience in the United Nations on each of these program thrusts. While there are isolated examples of modest success, the entire thrust must be termed a failure because the terms of trade have deteriorated, unemployment has increased, the poor masses are more destitute, a new class of rural poor has been created and debt burdens have risen astronomically in the developing countries.

Northern Chiefs' Position on Uranium: Regardless of our future conclusions about these unresolved questions, no further uranium development is acceptable to Indians until:

(a) land selection by Bands with unfulfilled treaty land entitlement is completed;

(b) the treaty rights of hunting, fishing, trapping and gathering are guaranteed against violation;

(c) the treaty rights for health, economic development and resource management are assured;

(d) we have the time and resources to carefully examine the many serious questions related to the uranium industry.

— Joint statement by the Meadow Lake and Prince Albert District Chiefs, October 7, 1977.

Let me give you but one example from **The Poverty Curtain** by Mahbub ul Haq, distinguished Pakistani economist and Director of the Policy Planning and Program Department of the World Bank. He writes (page 32):

> In Pakistan, which experienced a healthy growth rate during the 1960s, unemployment increased, real wages in the industrial sector declined by one-third, per capita income disparity... nearly doubled, and concentration of industrial wealth became an explosive economic and political issue. And in 1968, while the international world was still applauding Pakistan as a model of development, the system exploded — not only for political reasons but because of economic unrest. (By 1968 twenty-two family groups had acquired control of two-thirds of the industrial assets, 80 percent of banking and 70 percent of insurance in Pakistan.)

The fundamental reason why things have not improved for the majority of the poor countries of the world is that, do what they can, they do it within the framework of a global economic system where their position is pre-ordained as the suppliers of cheap labour and cheap raw materials. Virtually all the power lies with the multinational corporations and their home governments and there is no indication that this situation will change voluntarily.

There are two types of countries which differ from the norm. The first are countries like Taiwan and South Korea which have obtained massive influxes of capital from the U.S.A. (similar to the aid given to rebuild Western Europe through the Marshall Plan after World War II). China and Albania are examples of the second type. These latter, particularly China, initially cut themselves off from the multinational corporations and international financial institutions to ensure that their wealth was re-distributed internally.

New Policies for the Third World

Since the early sixties, third world leaders have been seriously re-analysing their experience and assessing their future. This has been done particularly through the United Nations Conference on Trade and Development (UNCTAD) and the Group of 77, a third world organization set up through UNCTAD to counter the power of the industrialized nations. Third world countries were primarily responsible for the proclamation of the New International Economic Order (NIEO) by the United Nations in 1975 and for several other subsequent initiatives by the United Nations.

It is fair to say that the NIEO challenges the global economic institutions and the apportionment of power in the global economic system. It asks for a more equitable trading and pricing structure, commodity agreements, control of multinational corporations, cancellation of debt, and an end to dumping of inappropriate technology, among other things.

There is a common assumption which still underlies most Western approaches to third world development, and that is that developing countries must somehow mimic the stages which have been taken by the industrialized countries. **It is precisely this which has been tried and which has failed.** But even by those still unconvinced, it must surely be recognized that it is now impossible for third world countries to follow the path of the industrialized countries. As I have pointed out, a major part of the resources on which the industrialized countries built their economies came from the third world. These were cheap – cheap labour, cheap raw materials **and** cheap energy. These are no longer cheap. Further, they are largely controlled by multinational corporations. Whose resources are the third world to exploit if they are to emulate this "development path" of the industrialized economies?

In any case why are we so arrogant as to think that third world countries wish to emulate us? Surely the declaration of the New International Economic Order tells us explicitly that they do not. This becomes more obvious when we look at the more specific program thrusts which have now been given priority by the third world and which have particular relevance to the question of energy needs. These program thrusts are:

1. Comprehensive rural development;
2. Self-reliance; and
3. Technical co-operation **among** developing countries.

There are others, such as the provision of decentralized basic health services, but the above three are probably most relevant to the subject we are discussing.

Comprehensive Rural Development

Most third world countries now place rural development as a first priority in national development strategy. They see a great deal of their present problem stemming from the mechanization and commercialization of agriculture based upon the Western model. This trend has speeded up the problems of urbanization and has created a new pool of rural poor while raising urban unemployment to frightening levels, sometimes as high as 50 percent.

The new thrust is to directly assist small farmers, improve the services available in rural communities and develop small scale industry for the provision of services to agriculture and for the processing of agricultural products primarily for domestic consumption.

This policy reflects the reality of life in the developing countries. First it is important to fully realize that two billion people — nearly half the world's population — still live in the rural areas of the third world. Their primary energy need is food. David Price of the International Development Research Centre (IDRC), in a letter to **Canadian Renewable Energy News**, December 78/January 79, points out that:

> The major area of fuel consumption is cooking for which the most common fuels are wood and cow dung. This represents as much as 80 percent of total fuel consumed in many countries.... In the agricultural sector,... the main energy input is human.... Next on the scale is animal power.

Nuclear electricity would be a most inappropriate way of providing the energy needs of developing countries facing these problems and realities. David Price points out the tremendous potential of increasing the efficiency of human and animal power by improved hand tools, bullock carts and plows. Clearly simple solar power appliances for cooking and crop drying would be a priority. The production of bio-gas would have a high potential in all developing countries. (Bio-gas for cooking and lighting is now used by 17 million people in Szechwan Province, China, alone.)

A related consideration is the capital cost of nuclear power installations. The current capital cost structure of conventional electric power generating systems are:

Thermal — $800 per installed kilowatt capacity;

Hydro — $1000 per installed kilowatt capacity;

Nuclear — variously estimated from $1200 to $2000 per installed kilowatt capacity (which **doesn't** include the many hidden costs I outlined earlier). A shortage of capital is one of the most serious limitations to development in the third world. This shortage is directly related to past dependence on imported technology inappropriate to their real developmental needs, through which huge debts have been incurred. In this situation, what possible rationale could justify the choice of the most capital-intensive, least labour-intensive technology?

The third consideration is the capacity of developing countries to absorb the relatively huge blocks of power provided by nuclear installations. The most economical size for nuclear power plants is in the 1000 to 1200 MW range. Further, one plant in an electrical power system should not be larger than fifteen percent of the system's capacity. Finally, a power distribution grid would be a precondition for such a system involving nuclear power. Very few developing countries have such grids or the capacity or need to accommodate large new blocks of power such as provided by nuclear installations.

Nuclear power is clearly inappropriate for the rural development aspirations and plans of developing countries from both the standpoint of it meeting their needs and their capacity to utilize it.

Self Reliance

In regard to the second program thrust — self-reliance — the developing countries are determined to utilize indigenous resources for their next period of development. On this count nuclear power is entirely inappropriate. The technology would have to be imported and financing the technology would place developing countries in ever greater debt. The fuel and its enrichment would have to be imported, supervison and servicing would come from the outside, and very few domestic jobs would become available. These matters were fully covered in the Saskatchewan Council for International Co-operation (SCIC) brief — "Nuclear Energy: Saskatchewan and the Third World" — presented to the Government in 1977.

More recent information on energy self-reliance comes from the United Nations Centre for Natural Resources, Energy and Transport:

> Governments of developing countries are sharply reducing the role multinational oil companies can play in the exploitation of petroleum resources.... More countries are assuming complete control over exploration within their own national boundaries, becoming sole owners of any oil found.... There is (an) increasing spread of knowledge about the workings of the international oil industry and, in particular, the beginning of the erosion of falsehoods and myths about the oil industry — myths that have been highly profitable to the companies and perpetrated by them and some international agencies in order to deter governments of underdeveloped countries from entering the oil industry. (**Development Forum,** November/December 1978)

From the standpoint of self-reliance it would be difficult to see how attempts to become more self-reliant in one energy source (oil) would be allowed to be undermined by developing countries permitting themselves to become more dependent on another energy source (nuclear power).

Technical Co-operation Among Developing Countries

The third development thrust of the third world — technical co-operation **among** developing countries — could be termed self-reliance on an international level. The third world countries are determined to use this method to break the economic stranglehold which the developed market economies have on them. They intend to increase their trade with each other, to share each other's technology (in place of inappropriate Western technology), to share manpower resources, to create commodity groupings (like OPEC) and to co-operate in other aspects of development.

A United Nations conference was held only this past summer (1978) in Argentina on the subject of Technical Co-operation among Developing Countries (TCDC). There were more countries represented at this U.N. conference than ever before. A firmer consensus emerged from this conference than ever before. There was hardly a word about this in the press but it was probably the most significant expression of unanimity about development ever achieved by the international community. The rules of the game are not going to change overnight as a result of this concensus but clearly a lot of new trends will be set in motion to begin to affect the imbalance between rich and poor countries.

The implications of TCDC for nuclear power development in the third world are clear. As nuclear power technology and industry is the sole property of the industrialized countries, it will not be acceptable as a general energy stategy for the developing countries. This will constitute a new, pervasive and progressive impediment to the further expansion of the nuclear industry into the third world.

As you well know, the peaceful use of atomic power has always been supported by the United Nations as a general principle. Even the Proclamation of the NIEO in 1975 endorsed the principle. TCDC does not question the principle but it does raise serious questions about the appropriateness of the technology and its control. This will effectively slow down the transfer of the technology whether or not nuclear power is considered an acceptable "peaceful use".

Responding to Requests

You may say: "Yes, but what if the government of country X requests a CANDU reactor: should we not respond to such a legitimate request?" This clearly presents a dilemma since past federal policy has put so much public money into this technology and so little into alternatives. However, there are a number of preconditions which must be a part of national external aid policy when such a question arises. First, is the technology safe, that is is it ethical to export it? Second, is the technology the most appropriate to the intended use and should alternatives be considered? Third, there is a responsibility to inform the interested customer about every aspect of the technology. (At the present time only the pro-nuclear side is given and this usually by nuclear industry salesmen and foreign affairs bureaucrats.)

The government should go further and determine, in some way, the level of public acceptance of the technology. I was in the Phillipines with the United Nations when the studies on nuclear power were conducted and the decision to proceed was made. The concerned public on Bataan Peninsula on the west coast of Luzon Island were not consulted in any manner: now they are vigorously opposing the construction of the plant and there is a real question whether it will ever be completed. Without public acceptance in the recipient country the exporting country is showing disrespect for and undermining the goal of self-determination abroad. It is thus a disservice to development.

Finally, as a matter of public policy, there should be no subsidies built in (above or below the table) which favour the particular technology or the purpose of which is to aid an ailing industry.

I very much suspect that present Canadian federal policy on this question is to accomplish the latter. CANDU is one of Canada's very few exportable hard technologies. It is important for crown corporation profits, for the employment of highly skilled people — many of whom it may be difficult to employ elsewhere — and for Canada's balance of payments situation. Yet the Canadian nuclear industry is in serious trouble. It can be assumed that the problems of the industry are at least **a** consideration, if not **the** consideration, of Canadian policy on this question. This may make good economic sense for the corporate economy, but it is hardly a positive world development position. This reality is made more difficult for the public to understand when you and other political leaders talk about "condemning the third world to a future of subsistence living" if nuclear power is not developed.

If we are as concerned as we say we are about the third world, should we not analyse their real needs and then respond in an appropriate manner? The immediate needs are simple needs: fuel for cooking, more efficient but simple agricultural hand and animal tools and bio-gas production. By satisfying these needs, human and other indigenous resources would be used more fully, food production would be increased, the quality of life would be vastly improved for the "poorest of the poor" and third world economies would be stabilized and made more self-reliant. This would provide the base for an appropriate and continuing development.

The research which is currently underway on Saskatchewan's sodium sulphate in regard to its use in solar power technology has far more significance for the developing countries than all the

---

It is not unfair to suggest that the government to a certain extent is involved in a conflict of interest. It wants to act safe, but I suspect it also wants to sell.

— Hon. Robert L. Stanfield,
former Leader of the Opposition.

---

uranium in the world. I commend it and trust that we will devote the same kind of effort and resources to developing it as the government is presently directing to uranium exploitation. If we want to prevent the greed that can come with riches, as Tommy Douglas appealed for in his talk to the last NDP convention banquet, then tying our development thrusts to the real needs of the third world, particularly to their energy needs, would be one of the surest ways of accomplishing this.

The development thrust which is needed in the world today can be identified by noting and reflecting upon some of the evident shortcomings of western industrialized societies. The common problems which are of greatest concern today are unemployment, inflation, inequitable distribution of income, increasing centralization and greater alienation. There are observable trends within society, if one cares to look for them, of not only concern for these questions but of actions to counteract them. I mention only a few: concern about environmental pollution and environmental health; concern about waste and obsolescence and the need for conservation of resources and recycling; concern about centralization of decision-making and the development of local community associations; concern about media control and advertising and the rise of consumer groups and an alternative press; back to the land initiatives and efforts to achieve greater self-reliance generally (such as insulating homes, home gardens, food co-ops, market gardens). There is a great stirring, largely spontaneous and unco-ordinated, all over the western world.

These concerns and the gradually emerging consensus about them are essentially the same as those of the third world. People everywhere are seeking ways of becoming more self-reliant and of achieving greater control over their own lives. They are looking for an appropriate, non-colonialist type of technology. They are

---

The hardest strokes of heaven fall in history upon those who imagine that they can control things in a sovereign manner, as though they were kings of the earth, playing Providence not only for themselves but for the far future — reaching out into the future with the wrong kind of farsightedness, and gambling on a lot of risky calculations in which there must never be a single mistake.

— Herbert Butterfield, *Christianity and History*.

---

looking for an ecologically inoffensive type of society. They no longer want to be masters of the world: they wish humanity to achieve a sensitive and sympathetic unity with the global ecosystem.

What we do in appropriate responses to third world needs has direct relevance in our own lives — although we may not be well aware of this as yet.

The Case of Japan

You made a specific reference to Japan in your letter. Clearly this is a unique situation. But to suggest that if we withhold uranium and nuclear power from Japan it will then be necessary for Japan to use fossil fuels and that will deny those fossil fuels to developing countries stretches credibility to the limit.

First, I would refer you to the evidence presented earlier which shows that there is no imminent shortage of fossil fuels. Thus the trade-off you suggest has no validity.

It is illuminating to review the historical development of Japan's energy regime. In his book **The Energy Crisis**, Michael Tanzer, now of the United Nations Centre for Natural Resources, Energy and Transport, wrote:

> Historically Japan... has lacked significant amounts of indigenous crude oil. Its energy economy ran primarily on coal and hydroelectric power, plus oil imported from the major international oil companies.... Japan's desire to obtain its own supply of crude oil was one of its principal motives in entering World War II.... After the war, with the American military occupation, the international oil companies took over with a vengeance. They turned Japan, like Western Europe, from a primarily coal-consuming country to an oil-based one, with oil's share of total energy rising from less than one-tenth in 1950 to over two-thirds in 1974. The companies collaborated with Japanese interests in building up a huge indigenous refinery industry since they were primarily interested in Japan as an outlet for their low-cost and high-profit oil. (page 87)

This example illustrates the overwhelming power of the multinational oil corporations, "The Big Seven", over national energy policies even in so-called presently developed countries. In the past few years these companies have extended their control of the energy sector of the global economy by obtaining large holdings in both coal and uranium supplies. (There are concrete examples of this in Saskatchewan's uranium — for example, Gulf Minerals, a subsidiary of Gulf Oil, in the Rabbit Lake uranium mine.) It can be assumed, therefore, that pricing and other arrangements will be used to maintain an artificial balance between the prices of oil, gas, coal and uranium to ensure maximum use of energy resources and thus maximum profits.

In this context it is difficult to see how the withholding of uranium from Japan will somehow affect energy supplies available for the third world. If a profit is to be made from supplying energy, anywhere in the world, the energy multinationals will see that energy, in some form, will be there. If the country is very naive, or very corrupt, or anxious to get into nuclear weaponry, the energy companies will choose those places, with the connivance of their home governments, to dump nuclear power plants.

Underlying Considerations

Aside from the profit motive, which is the main dynamo of the global economic system (rather than the satisfaction of basic human needs), there are two other current preoccupations of western industrialized countries in regard to energy supplies. One, to which I referred earlier, is the unhappy state of the nuclear plant manufacturing and construction industry. This I believe to be a prime motivation behind the policies of western nuclear nations. As it is becoming more difficult to sell nuclear plants at home we are making supreme efforts to sell them abroad — to third world countries — to save the industry! (Please see again the **Business Week** Special Report, 25 December, 1978.)

But the underlying concern of the home governments of the energy multinationals, particularly the U.S. and western Europe, is **security** of supplies. These governments are still dependent on OPEC oil, particularly that of Saudi Arabia and the Persian Gulf. They are therefore determined to phase in nuclear power as quickly as possible because uranium supplies are indigenous to the western, white, industrialized economies (Canada, U.S.A., Australia and South Africa contain the major reserves of uranium).

The concern about security of energy supplies is a legitimate concern of the industrialized countries. The current situation in Iran is an example of how sensitive our sources of supply are to domestic unrest in countries with repressive regimes. However, if this problem is to be solved, it must be clearly identified as the problem. There is a difference between the type of policies required to meet the problem of **short** supplies and the problem of **security** of supplies. If supplies are short, then there is no alternative to programs of conservation and the development of renewable sources of energy. If not, policy has to search for the reasons why security of supplies is in doubt and to find ways of dealing with those problems. Certainly, in the Middle East it cannot be obtained by huge arms sales and neglect of local social and poverty problems. All of our efforts should be devoted towards support for creating equity and peace among our suppliers and their neighbours. This the Western world is not doing because there are no short-run profits to be made. But, by not attacking the roots of the basic problem of security of supply, and recognizing that as the problem, we also do our own economies harm in the long term because of increasing global disparity and insecurity.

Summary

The scenario which the multinational energy corporations and their home governments are following could be summarized as follows. Huge energy-consuming economies have been established. It is profitable to maintain these. However, traditional supply sources are insecure. There are other supplies — uranium and coal — which seem more secure. These are substantially controlled by the same energy corporations, but the public is reticent about their development — particularly that of uranium. So two things are done. First, the public is frightened about freezing in the dark by developing and spreading the myth of an overall energy shortage. Second, nuclear technology and industry is kept alive by selling it at rock bottom prices to third world countries — in the hope that public opposition at home can be stilled.

The real needs and aspirations of the third world are not a part of this corporate scenario in any respect. In fact by continuing the Madison Avenue type of selling of nuclear power to the third world, the industrialized west is doing a most serious disservice to the third world's developmental goals and processes. The absence of nuclear power does not "condemn them to a future of subsistence living". Rather its presence would lock third world countries more securely into dependency upon the global economic system controlled by multinational corporations and the western industrialized nations. This is the essence of imperialism and neo-colonialism.

I think I need say nothing more about the lack of morality of the industrialized countries on this question. The corporations and governments of these countries are at present building their economic self-reliance by eroding the efforts of the third world to become more self-reliant.

4. Nuclear Proliferation

> **You point out** that other forms of research and production besides nuclear power have serious weapons spinoffs and that there is "no cry to stop those forms of endeavour". You also note that "there are weapons enough available now to destroy the world totally" and that the military balance of terror "has long since been tipped". You talk of my "dark view of human nature" on this question and you hope "that the great military machines can be stopped by our collective will". (paragraph 7)

I agree there are many imminent threats to world peace and that every effort should be made to stop all forms of weaponry development. I believe our view of whether or not the balance has been tipped is probably largely semantic.

But all of this is merely an idealistic statement. What we require are concrete, practical acts to help accomplish the goal of world disarmament and peace. It is at the level of practicality that the issue of uranium and nuclear proliferation must be addressed. I will deal with your comments in that light.

Weaponry Related Research

Other forms of research leading to weaponry development

which you refer to are medical and space research. There is a great deal of difference between these and nuclear weapons. In regard to medical research, there are already international covenants on biological warfare and we know that some stocks of biological warfare supplies have already been destroyed. Furthermore only a very small component of medical research has direct relevance to weaponry development.

In the case of space research I would suggest that this is presently dominated by the two superpowers, as was originally the case with nuclear weapons technology. This field is closely related to nuclear warfare (shooting down satellites and missiles, etc.) and will undoubtedly be one of the aspects of discussion in the ongoing SALT talks.

However, the fundamental weakness of your argument here is that you are comparing unlikes. You talk about the weapons spinoff from research. But that is not the state of nuclear weaponry. It was a **direct result** of research designed to produce weapons (not just a spinoff) but, even more important, plutonium, the fuel for these nuclear weapons, is a by-product of the nuclear power industry. That is why it is a more urgent and imminent threat than the others you mention: it is a problem of a highly different quality and magnitude.

Safeguards

I am aware that there is international inspection of nuclear technology and that Canada has further safeguards against misuse. But the experience with India, which after buying a CANDU reactor, used the spent fuel to build and test a nuclear bomb, demonstrates the real effectiveness of these safeguards.

In regard to safeguards through the International Atomic Energy Agency (IAEA) there has been growing concern for some years about the capacity of the agency to fulfill its inspection responsibility. A memorandum on this subject from the Director of the Safeguards and Inspection Division of IAEA, Sloboden Nakicenovic from Yugoslavia, was leaked during the June 1978 UN Disarmament Conference. Director Nakicenovic claimed there is underrepresentation of non-aligned countries within the IAEA; that the inspection procedures at bulk fuel fabrication and reprocessing plants are insufficient; that the data processing system needed to analyze inspection results is inadequate; and that a suitable method for assessing the quantities of bomb-grade nuclear materials now under safeguards is lacking. He claimed that the Agency's common practice of abiding by the interests of "influential" nuclear exporting nations has not only jeopardized the enforcement of safeguards but has also eroded world confidence in the organization's fairness and effectiveness. (See **Not Man Apart**, Friends of the Earth, July/August, 1978.)

I can support this judgment of the adequacy of IAEA inspection services from my own experience with the United Nations. In one country in which I was the Resident Representative, some of the inspections of that country's research reactor were conducted by an IAEA staff member who was a national of the very country he was inspecting and who was dependent on his government for a continuation of his employment with the IAEA.

A situation like this is hardly surprising because the IAEA is fully committed to nuclear power development. That is its mandate. Like Atomic Energy of Canada Limited, it is wholly pro-nuclear. There is no room for a debate about any of the problems of nuclear power within such organizations and their inspection and regulatory functions reflect the committed pro-nuclear stance of the organization. Clearly such regulatory responsibilities are irrevocably weakened by a fundamental conflict of interest.

It is not helpful to sidestep these realities about the inadequacy of safeguards in making local, provincial or national decisions. It is clear that, within the present state of national and international safeguards, **the potential for weapons production goes along with every extension of nuclear technology and every extension of uranium mining.** When Saskatchewan makes a decision on this matter it must fully recognize the present realities about safeguards and the responsibility which flows from those realities.

Realities about Proliferation

It is useful to review some of the statements of third world spokesmen on this issue. After India exploded her bomb, former Prime Minister Bhutto of Pakistan said he will "never surrender to any nuclear blackmail by India. The people of Pakistan are ready to offer any sacrifice, and even to eat grass, to ensure nuclear parity with India."

When Argentina was negotiating the agreement with Canada for nuclear technology, an Argentina spokesman was asked to comment on the conditions of the agreement. He said: "It's really a little silly. We'll sign the agreement all right. But how do they expect to enforce it? Besides, we wouldn't dream of building a nuclear bomb — unless Brazil does!"

Can there be any real doubt but that nuclear weaponry is at least one of the considerations when a government decides to build a nuclear reactor? How else can you explain India's action in view of her huge coal reserves? Her perceived enemy was China which had the bomb. And India's "achievement" brought Pakistan into the Club. How else can you explain Brazil, which has enough water power to supply all her electricity needs until the next century, being interested in nuclear power? And Brazil's decision brought in Argentina. How else can you explain Iran, which sits on top of ten percent of the world's proven reserves of oil? **It is stretching credibility to the limit to believe that the option of weaponry is not an objective of most nations which now have, or want, nuclear technology.**

In the case of Brazil, West Gemany supplied the technology with safeguards which the U.S. government considers inadequate. Brazil is now having second thoughts about the $20 billion deal with West Germany. Brazilian officials found that electricity generated from the nuclear plants would cost three times as much as hydroelectric power. They now say they may only want two of the proposed eight plants. (See **Seven Days**, 27 October 1978)

Most informed estimates tell us there are now 12 countries which have nuclear weapons. In addition to the big five, this includes Israel, Argentina, South Africa, South Korea, Spain, Taiwan and India: not a reassuring group, I am sure you will agree. Proliferation is a fact but I find it hard to understand why we shouldn't be seriously concerned about further proliferation.

Your dismissal of any responsibility for the total problem of nuclear arms is, of course, quite typical of the non-responsibility claimed by every government that is contributing to the total problem. No one is totally responsible, therefore each contributor to the problem can disclaim responsibility! What a perversion of logic! It is similar to the way levels of government presently pass the buck about the problems of native Canadians, and is equally counter-productive.

One of the most serious new possibilities for further proliferation was unveiled by the U.S. General Accounting Office (GAO) in October 1978. The GAO says it may be possible for many countries without nuclear weapons to secretly build and operate plants for making weapons-grade plutonium from spent atomic power plant fuel. Such a plant could produce about 11 pounds of plutonium daily. The GAO said this finding increases the importance of efforts to prevent diversion of nuclear fuel materials into weapons systems. (Regina **Leader-Post**, AP — Washington, 25 October 1978)

Black Market Developments

Probably the most serious aspect of the proliferation question is the growing evidence of illegal trade and black market operations in weapons-grade material. This is partly due to the inadequacy of safeguards discussed earlier. First indications of actual black market operations came from a 1976 GAO report that government-owned nuclear facilities in the U.S. could not account for 11,000 pounds of plutonium and enriched uranium. A second investigation of privately owned facilities reported another 15,000 to 20,000 pounds could not be accounted for. (It takes only 12 pounds of plutonium to produce a bomb.)

This was first revealed in the U.S. journal **Rolling Stone**, noted for its investigative reporting, in its May 19, 1977 issue. Associated Press released the story several months later. Excerpts from the **Rolling Stone** report follow:

> Under normal processing conditions some plutonium is inevitably lost in pipes and industrial machinery. But the rest, according to Government and industry people familiar with the situation is being stockpiled for sale or has already been diverted for sale . . .
>
> In the past three years. . . a reckless and disorderly nuclear market has sprung up, essentially independent of the U.S. Government. Circumstantial evidence suggests the presence of black marketeers or nuclear pirates, though not enough is known yet to discuss their role. But also prospering are "gray marketeers" who take advantage of legal loopholes.

The report then goes on to describe two U.S.-based nuclear brokerages which were established after 1973 when the Nixon Administration ruled that U.S. companies could trade in nuclear commoditites.

> Two. . . brokerages were established: the World Nuclear Fuel Market (WNFM) based in Atlanta,

> Georgia and the Separative Work Unit Corporation (SWUCO) with headquarters in Gaithersburg, Maryland... WNFM is set up like a consortium, with 79 nuclear companies and utilities as members: 38 from the U.S. and 41 from 17 other nations... WNFM has disseminated 112 pounds of plutonium, enough for nine bombs, in two separate transactions. One shipment, for 35 pounds, was sent from a U.S. company to a foreign buyer. WNFM officials refuse to say where the plutonium went.

In February 1978, Associated Press released a story from Washington which dealt with the loss from just one of the U.S. nuclear installations:

> The top staff official of the Nuclear Regulatory Commission (NRC) acknowledged Monday for the first time that enriched uranium (as much as 200 pounds) may have been stolen from an Apollo, Pennsylvania nuclear plant in the mid-60's....
>
> For years the Government had publicly insisted the disappearance was due to an accounting error. But recent documents suggest a growing feeling within the Government that the material was stolen. (Regina **Leader-Post**, 28 February, 1978)

Seven months later, in August 1978, another Washington AP dispatch followed up on this story:

> Officials from both the Energy Research and Development Administration (ERDA) and the Nuclear Regulatory Commission (NRC) concede the Apollo plant has been unable to trace 381 pounds of enriched uranium found to be unaccounted for in 1965...
>
> A Congressional source said Friday the CIA has "really overwhelming" circumstantial evidence that the nuclear material went to Israel...
>
> The CIA is reported to have reached its conclusion after investigating relationships and activities involving the plant.... The source said the spy agency's evidence includes patterns of travel to Israel by officers and employees of the company... plus a corporate connection to an Israeli firm. (Regina **Leader-Post**, 22 August, 1978)

Another example of the proliferation threat from uranium is the case of the ship carrying 200 tons of uranium ore from Antwerp to Genoa in November 1968 which completely disappeared. It was discovered some weeks later: under a new name and a new registry — but with no uranium aboard. Intelligence officials from the CIA and three European countries believe this uranium also ended up in Israel. If it did and if it was run through the kind of reactor Israel has had since 1963 it would yield enough enriched uranium to produce 135 A-bombs.

I am concerned about the proliferation of all kinds of weaponry. However, when an economy begins to be based on an energy fuel cycle which actually produces weapons-grade material as a by-product, and that by-product can command a high price, it is inevitable that a black market will develop which will proliferate the product to irresponsible regimes and criminal elements. This is a far greater present danger than exists with any other technology and there is already sufficient evidence that regulations and policing are not controlling it and cannot control it.

This concern is not based on a "dark" view of human nature. It is based upon a realistic analysis of Western energy policy and of some very concrete results of that policy including the lack of preventive foresight regarding proliferation and the increasing risks of nuclear war. Avoiding these problems by abstract philosophizing, as you have done, is hardly an answer.

These weapons will soon fall into many hands in many corners of the world — with incalculable consequences for us all. This danger is the direct result of the ***uncontrolled growth of the nuclear power industry,*** *which is making widely available the materials needed for such weapons. The peoples of the world must recognize the danger of what is going on and act to* ***protect this and future generations.***

— Herbert Scoville, speaking for himself and Drs. George Kistiakowsky, Theodore Taylor, George Rathjens, and Bernard Feld, Granada Television, 1976.

Stopping the Military Machines

You express your hope that "the great military machines can be stopped by our collective will". There definitely must be a great outpouring of concern from the public on these matters: this is a question on which I continue to be personally involved. However we must be clear about what we mean when we talk of "collective will". Surely the precondition of an expression of collective will is an informed public and responsible leadership on the issue.

If this is so, then I maintain that the Government has not fulfilled its obligation to properly inform the public on the broad issues connected with uranium mining. The time allowed for the Bayda Inquiry was shamefully short. All native, labour and public interest groups pleaded for an extension of the time: it was denied. The Government did provide some money to help these groups to participate in the Inquiry: however, it was very small compared with the money spent by the pro-nuclear corporations and government agencies. In any case it did not resolve the problems of the short time available to inform and involve the public and the narrowly conceived and interpreted terms of reference.

The Inquiry sponsored public debates throughout the province. These were held in the middle of the summer: the worst possible time for Saskatchewan people. I attended ten of these as the non-nuclear spokesperson. The average attendance was less than twenty people!

The media did provide reasonable good local coverage but this is no credit to Government. Further, media coverage is very much after the event: it rarely assists in developing effective

---

... the majority ought to be sovereign, rather than the few. For it is possible that the many, no one of whom taken singly is a good man, may yet taken all together be better than the few not individually but collectively, ...

— Aristotle

---

public understanding and involvement. Since the Bayda Report, the Saskatchewan media has been constantly reinforcing the uranium boom mentality, which tends to take attention off these underlying, fundamental questions.

Finally the Government refused to share the report with the public until the Government had decided its position. You took every possible precaution to ensure that the public did not have access to the Report: even to the extent of locking the press into a room with copies of the report until the time of the press conference.

This was indeed a strange way to generate a "collective will" about an issue which I have shown has a clear and direct connection with the threat of war. You expressed the hope "that the great military machines can be stopped by our collective will" but when it comes to practical steps towards developing that collective will your actions belie your words.

Your summary paragraph on this point (paragraph 8) reads as follows:

> We should not prevent the beneficial peaceful uses of uranium simply because uranium is a component of nuclear weapons. The prevention of nuclear war lies in the political and moral sphere and will not be affected one way or another, I believe, by a ban on uranium mining.

I have shown the direct connection between the so-called peaceful uses and the weapons potential of uranium. However, in view of your own personal experience on this matter, I find it difficult to understand how you can express such a naive belief. Let me remind you of that personal connection.

In early 1976, you visited South Korea, reportedly, among other things, to discuss sales of Saskatchewan uranium to that country. You did not know at the time that the South Korean government had decided to produce nuclear weapons. This decision was unanimously approved by a South Korean weapons exploitation committee in the early 1970s. However, the South Korean nuclear bomb plan ran into problems in 1975 due to the difficulty of getting nuclear fuels. You arrive on the scene one year later, innocently, to sell Saskatchewan uranium to save the South Korean people from disaster, while your clients were looking for uranium for bombs!

Summary

What you seem to be saying on this subject is that these problems are too big for us and that it therefore doesn't matter whether or not we sell uranium on the world market. You overlook an important fact. The reserves of uranium in northern Saskatchewan probably account for ten percent of world supplies (at the rate exploration is proceeding it will probably soon exceed this percentage). However, total known world reserves are insufficient to fuel a nuclear industry of even one-half the size of that now projected by the pro-nuclear forces. Therefore, holding back our supplies would have a profound effect on the growth of this most dangerous technology. It would be a political and moral decision of great historical significance for mankind. (See **Nuclear Power: The Unviable Option**, John J. Berger, 1976, Chapter 7).

A real assessment of the proliferation problem shows conclusively that any support of the nuclear industry at this point in history adds significantly to the threat of war. As I said earlier, in view of this reality, what the world requires are concrete, practical acts to help accomplish the goals of safe renewable energy sources and disarmament and peace. Statements of ideals are of nothing without the practical acts. Saskatchewan's practical act is clear: **we should leave uranium in the ground at this stage of history.**

5. Radiation Standards and Public Health

> **You acknowledge** the dramatic changes which have taken place in scientific opinion recently. You note that your standards for uranium mining are "the toughest in North America" and that you are prepared to change them if further research found them to be inadequate. Then you note that risk is always with us and state that "uranium mining is less risky than coal mining or asbestos mining". Finally you state, "We do not believe the risk is sufficient to stop the mining." (paragraph 9)

I fail to see the purpose of your first general point about science and scientists. If you are obliquely saying that you must place your faith on these matters in some elitist scientific or "expert" group then I must take strong exception.

Science adds knowledge and increases understanding of the universe, the world, society and all of the components and interrelationships involved. Science, even in this classical sense, has never been "pure". It has always been, to some extent, the handmaiden of power groups within society. In the present period of history it has acquired two important and related characteristics. First, it has become increasingly more highly specialized and compartmentalized. This means that the scientist lives and works in a constantly narrowing area of specialization. He therefore becomes less relevant to the development of public policy which must always be multi-faceted and increasingly so.

The other characteristic of modern day "science" is the increasing control of research by industry, the military and governments. Science, in the ideal sense, has lost much of its freedom of scientific investigation simply through the manner in which research priorities and funding are determined. Consequently it has become evermore the servant of centralized vested interests. Many of those referred to as "scientists" are actually nothing more than technicians who play a particular role in maintaining a status-quo approach in industry or government.

If the balance between science and technology is shifted in favor of the latter, as I believe has happened in the nuclear field, then the technology virtually runs amok (not to be confused with Amok!). Unless that accumulating technology in all its ramifications receives continuing research attention, there is no way of judging the effect of that technology on society or nature. Modern day corporate society is noted for its dependence on technology and its unwillingness to provide science with the resources to properly establish baseline data and constantly assess the effects of technological developments.

Nowhere are these characteristics seen more clearly than in the field of nuclear technology. Einstein created the understanding which permitted the technology to proceed (he also warned of its serious dangers). The bulk of the effort since that time has been on the technology, first for weapons and subsequently for nuclear power. Tens of billions of dollars, mostly public money, has been spent on the development of the technology and the training of technologists. Society now has a cadre of these people, who some call scientists, but who are primarily technicians trained for a particular industry and a particular type of economic organization.

Fortunately, there are a growing number of people working in the nuclear field who have tried to maintain their integrity and the public's trust. They have proceeded with their scientific research in the spirit of open inquiry, freedom of information and public accountability. They have raised their voices in protest at some of the weaknesses they have seen in nuclear technology.

Many have lost or quit their jobs. Increasingly they find funding for their research cut off because of the serious questions it raises about the technology.

I raise this point because the Bayda Inquiry, on which your Government depends so fully, demonstrates a virtually complete dependence upon the advice of the narrowly defined "nuclear scientist". It looked kindly upon those persons who work in various segments of the nuclear industry, overlooking the fact that these persons would be less than completely objective in their views because of their dependence on the industry for their livelihood and their tunnel-vision about the larger questions. In the Preface to its Report, the Inquiry says: "The proponents count most of the pre-eminent nuclear scientists" (page 11). That statement, of course, could be seriously questioned: it is, however, a clear indication of how the Board felt about the high priests of the technology under question.

The term "nuclear scientist" has not kept up with the real growth of scientific knowledge in this field. It is still limited generally to a nuclear physicist who understands the fission process as it is used for electricity and weapons production. However, other branches of science, particularly in the health field, have developed since the advent of nuclear technology. The fields of health physics and radiation epidemiology are probably the most important. And it was in these very areas where the Bayda Inquiry was most wanting, simply because the members of the Inquiry were only willing to listen to nuclear scientists in the narrow sense (i.e. technologists) and were unwilling to listen to what science in the broadest sense was trying to tell them.

I want to describe only a few of the Bayda Report's shortcomings in the manner in which it dealt with the critical health questions. Further detailed examples will be provided in our overall critique of the Report.

Radiation Standards

The Bayda Inquiry totally evaded, in a scientific sense, an evaluation of existing radiation standards, although its Terms of Reference clearly called for such an evaluation. The first two sentences of the section of its report dealing with radiation standards show the bias on this question:

> The emission of radiation is associated with many activities from which society derives **benefits** (my

> emphasis). To gain those benefits man has to expose himself to some radiation... (paragraph 3.33, page 42)

This is hardly the way to begin an objective, scientific study into radiation standards. From a public health perspective, would it not be wiser to start with the premise that "the emission of radiation is associated with many activities for which society pays costs"? The discussion by Bayda was prejudiced at the outset by the appeal to the **benefits** of radiation-associated activities.

The Bayda Report does not discuss the methods by which existing standards were calculated. It simply appeals to the "scientific reputations and independence (paragraph 3.35, page 43) of bodies like the International Commission on Radiation Protection (ICRP).

When one looks at some of the recent evidence about radiation exposure and health, the weakness of Bayda's recommendations can be seen. For example, Dr. Karl Morgan, himself a former chairman of the ICRP and a pioneer in the field of health physics, concludes from his study of recent literature in the field: "There is no safe level of exposure and there is no dose of radiation so low that the risk of malignancy is zero." He considers the most recent report of the ICRP (1977) to be "a retrograde step... because it comes at a time when their own internal reports emphasize that the cancer risk is many times what we considered it to be 15 years ago." He recommends the permitted radiation level be cut in half and admits he would argue for an even further reduction except for the difficulty this would create for the nuclear industry. (**Bulletin of Atomic Scientists**, September, 1978)

Dr. Joseph Rotblat, Emeritus Professor of Physics at the University of London and a past president of the British Institute of Radiology also says that "new knowledge indicates that the carcinogenic action of radiation is much greater than we thought earlier". He goes further than Dr. Morgan and concludes that "a reduction of the dose limit by a factor of five would be justified". Dr. Rotblat's article in the same issue of the **Bulletin of Atomic Scientists** continues:

> It has been suggested by Morgan that a reduction of the dose limit by more than a factor of two is not acceptable because of the impact it would have on the nuclear industry. This argument seems illogical to me. If dose limits were to be decided by what a given industry can afford, then we might as well abandon all

> attempts at estimating risk factors from actual data, and leave it to the economy to set radiation levels. These would then differ from country to country, and even from industry to industry. Once we move away from scientific criteria, the situation becomes untenable.

Dr. David Bates of the Faculty of Medicine, University of British Columbia, presented a paper on "Occupational and Environmental Health Considerations" at a Nuclear Policy Conference sponsored by Carleton University in November, 1978 which I also attended. He was quite critical of the Bayda Report in the narrow and superficial manner in which the occupational health issue was dealt with, especially the lack of consideration of the effects of internal radiation in the calculation of the health risks to workers. In discussion, Dr. Bates appealed to colleagues to give more serious attention to the recent and rapidly accumulating body of research on the dangers of low-level radiation.

Bayda's rush to accept the ICRP recommendations and to overlook the serious questions of such eminent persons in the field of health physics as Doctors Morgan, Rotblat, Bertell, Mancuso, (and many others) is indicative of his desire to sidestep serious consideration of the adequacy of radiation standards. Bayda overlooks the fact that advances in human knowledge are continually being made and these continue to raise questions about edicts handed down by organizations like ICRP.

You say in your letter that if further research show that regulatory standards are inadequate you are prepared to change them. The further evidence was available when the Bayda Inquiry was being held. **I now challenge you, therefore, to change the standards as you have promised.** I recommend that you bring in these eminent scientists and hear what they have to say in an open forum.

The Bias of the Bayda Inquiry

There was not only a lack of proper investigation of health matters by the Bayda group; there was outright bias in its treatment of evidence on these matters. I give but one example.

The Bayda Inquiry was informed about a study of nuclear plant workers at Hanford, Washington, which found that cancer deaths rates were much higher than expected at radiation levels well below the existing standard. The Inquiry dismissed this study by saying, "However, the procedures used in the statistical analysis have been severly criticized" (paragraph 3.64, page 52).

Bayda does not mention the nature of this dispute. He does not report the fact that the authors replied to the criticisms at an International Atomic Energy Agency meeting and reconfirmed their original findings in a reanalysis of the data.

The researcher involved in the controversy was Dr. Alice Stewart, a pioneer of radiation epidemiology with more than 20 years experience in the field. In the Bayda hearings, Dr. Nenot of the French National Power Company, Electricite de France, an Amok witness, criticized her work. This testimony was entered as formal evidence. Dr. Nenot claimed that he knew Dr. Stewart and then made patronizing remarks about her being a young, clever girl but not scientifically serious. He then said she was redoing the study, leaving the impression she had abandoned it under criticism. Dr. Stewart subsequently submitted a letter to the Board of Inquiry which states in part:

> I do not think I have ever met or corresponded with Dr. Nenot and his claim that he knows me well and has good friends who work with me comes as a great surprise. It is also surprising to find a woman of my age (she was 76 in 1978) being described as a clever little girl who, scientifically speaking, is not completely serious. There have, however, been several occasions when scientists who favour rapid development of the nuclear industry have used similar phrases for the express purpose of undermining confidence in the Kneale analysis of Hanford data.
>
> According to Dr. Nenot he is familiar with this work. However, his criticisms are so wide of the mark that George Kneale and I have difficulty in accepting this statement or attempting any serious rebuttal.

The Bayda Report accepts the clear misrepresentation of the French bureaucrat, who acknowledges in his testimony that radiation epidemiology ". . . is difficult to explain because I don't understand it quite well, I am not a specialist." (**Transcript**, page 2503). It dismisses the evidence of a renowned British scientist.

Asbestos and Uranium

You raise the question of comparative risks in asbestos mining.

I personally favor the reduction of occupational and public health risks in every part of industry. That is the essence of a preventive and environmental approach to health. Arguing about which is most risky is a non-productive argument which can easily lose sight of this goal. Our subject of discussion is the risk of uranium mining and nuclear power. This risk can not be considered in the abstract: it must be considered in the real world which already includes high levels of industrial pollution.

However, the subject of asbestos is moot because, with its longer history, it may well tell us something about how government and industry will be treating Saskatchewan uranium workers in the future.

Insurance companies refused to grant life insurance to asbestos workers because of health hazards of this industry as far back as the 1920s. Epidemiologists have shown the direct relationship between asbestos and cancer and other illnesses for many years. However, government compensation boards and mining companies stubbornly refused to acknowledge the connection until very recently. In its fight to avoid responsibility, the industry, with the support of government and some consultants from academia, used every research gimmick and methodology available to them to distort and disguise the facts on this question. (Please see the article, "Scientists not Neutral" from **Commonwealth**, 24 January, 1978). Finally, through public pressure, the regulations have been changed, after years of suffering and privation, and deaths of thousands of workers.

The cursory examination of health issues by Bayda does not bode well for the workers (or their offspring) you are hoping to entice into the uranium mining industry.

Furthermore, no provision has been made for base-line studies

---

When society does decide to risk life for a comparable life gain, then the individual must have some way of opting out of the risk. For example, one can choose not to travel in a plane or not to live near a fossil fuel plant. Generally polluting air, land and water with radioactive material, undetectable by human senses, is an unacceptable risk from which the more sensitive segment of society has no protection."

— Rosalie Bertell, Ph.D., GNSH
Christianity and Crisis, Oct. 16, 1978.

---

which would provide at least some assurance to the workers who will develop illnesses that there will be some adequate scientific basis upon which to base compensation claims in the future. It is noteworthy that even though the baseline research on other species was found wanting, the Bayda Inquiry did go after such information for birds, animals and plants but **not** for humans.

Summary

You state: We do not believe the risk is sufficient to stop the mining.'' The risk for whom? The Cabinet which made the decision? Amok's shareholders? Or the northern workers? Is there any way of striking an average ''risk'' when individual lives are involved — and not only the life of this generation but of generations to come?

Under **The Occupational Health and Safety Act**, 1977, passed by your government, workers have the right to leave an unsafe workplace. Are you and Amok going to have the workers fully informed by independent health physics authorities about the real and potential risks of radiation which are undetectable to human senses? Are you going to provide them with alternative employment opportunities if they decide the risk is too great? Or, because of lack of other opportunities, are they going to be forced into these high-risk jobs and take their increased chance of cancer 20 years down the road?

There is a growing body of evidence, with a long data base, which is converging on the conclusion that health hazards from radiation are perhaps twenty times greater than was thought at the time present radiation standards were set. This evidence says that even your government's standards, which you say are ''the toughest in North America'' are seriously deficient. Bayda's superficial treatment of the subject is seriously negligent. The fact that there is such a serious dispute in the scientific community over the health hazards of low-level radiation should

---

Radioactivity causes mutations in the structure of DNA, the long molecule that contains the coded genetic information necessary for the development of a human being. That is perhaps its most devious effect, since the damage may not appear for a generation or more.

— The Cousteau Society

---

certainly provide adequate reason for pause. **This unresolved question of human and environmental health is surely a compelling reason for a moratorium.** Otherwise people will be justified in asking: is the role of government to protect the health and well-being of its citizens? Or is it primarily to assist industry to exploit resources?

Only you can answer that question.

6. Nuclear Waste Storage

> **You state:** "On the issue of radioactive waste disposal we have simply had **to make a leap of faith** (my emphasis) and assume that a satisfactory means of disposal will shortly be found. The waste created by nuclear power generation is only a tiny fraction of the total problem, as you know, so it did not seem to us very useful to stop the generation of nuclear power when that gesture would not appreciably diminish the problem of nuclear waste." (paragraph 10)

I am appalled to learn that you could make such a statement. These wastes are highly poisonous. They will last for untold generations — 250,000 years or more. And yet, in the face of a problem of such immensity, you can say that you "simply had to make a leap of faith and assume that a satisfactory means of disposal will shortly be found."

You admit that there is, at present, no known method of nuclear waste disposal. Atomic Energy of Canada Limited (AECL) assures us it will have a means of disposal by 1990. But, as **Business Week**'s Special Report of December 25, 1978 tells us: "despite decades of government assurances that safe disposal is technically possible, the projected startup date for a permanent repository is further away today than it was five years ago."

In the interest of clarity on this question, let us recognize that there is no such thing as "disposal" of nuclear wastes. It is important to note that this waste is a product of the technology. Only the processes of nature, over many thousands of years, will dispose (that is, deal conclusively with) these wastes. **All man and his technology can do is store nuclear wastes: we cannot dis-**

pose of them.

What the waste disposal system might be, whether it will be safe, what the direct and indirect costs will be, cannot be answered until the industry develops the technology. Clearly a moratorium on the nuclear industry is the only morally responsible approach in the absence of any existing technology for this critical part of the nuclear fuel cycle.

Dealing with a critical question like this "on faith" is simply placing an unjustified confidence in technology. It is typical of the increasing arrogance of centralized industry and government. It places the general public welfare subservient to a misplaced confidence in unproven technological "fixes". Some people refer to this, not as science, which is based on rational skepticism, but a new religion of technological priests.

I must confess that I simply do not understand your second sentence. If you are saying that the waste from nuclear power

---

Men of high position are allowed, by a special act of grace, to accommodate their reasoning to the answer they need. Logic is required only in those of lesser rank.

— John Kenneth Galbraith

---

plants is not as dangerous as the waste from mining, or processing, or fuel enrichment, then I think you are missing some essential points.

At the present time, the wastes of greatest concern to the industry are the wastes from nuclear power plants. There is an urgency about this because storage capacity at nuclear plants is rapidly filling up. The **Business Week** Special Report says: "Some commercial reactors (in the U.S.) face shutdowns in the 1980s if additional temporary storage is not built."

It is only recently that concern has grown about the environmental health problems related to mine tailings. This came to public attention when highly dangerous radon gas emissions were located in U.S. and Canadian centres (such as Port Hope, Ontario and Uranium City in northern Saskatchewan) where mine tailings had been used for landfill.

A recent publication, "Natural Radioactivity Contamination Problem 1978" of the U.S. Environmental Protection Agency provides important information on this subject:

> While there are over 100 naturally occurring

> radionuclides, **public health problems are usually limited to 30 or more radionuclides in the uranium and thorium decay series** because of their relative abundance and toxicity.... Other examples of the increased population exposure... include the **radon problems** in several Western States **due to the use of radioactive tailings**... (my emphasis).

Another report from the same Agency, "Radiological Quality of the Environment, 1976", estimated from all available data that the largest radiation dose for individuals, received from all sources, is derived from radioactive material added to the environment, primarily due to the growth of nuclear technology.

A statement such as you make shows a lack of understanding of the pervasive nature of the nuclear waste problem. You overlook the holistic nature of the nuclear fuel cycle. Waste is a product of the entire cycle of mining, milling, transporting, refining, fuel enrichment, use of the fuel for production of electricity and weapons, reprocessing and reuse of the fuel. To look at waste as a product of one or more parts of this cycle and to indicate that one or the other is more important misses the point that waste and its disposal (storage) is a reality once you have engaged in the cycle at any point. It is a complete cycle and it is global in nature. Any enterprise which enters that global cycle at any point has a moral responsibility to ensure that the total waste disposal problem of the total global cycle has a solution before entering into it. If only for selfish and/or provincial reasons such an internationalist perspective and commitment is necessary.

There is a real urgency to stopping nuclear energy production because there is not **and cannot be** a safe program for **disposing** of radioactive wastes. The reasons are basically simple, do not depend on any complicated scientific arguments, and cannot be refuted or made irrelevent by any conceivable increase in scientific knowledge or technological capacity. To make this point, I give below quotations from the report, "Considerations of Environmental Protection Criteria for Radioactive Waste", U.S. Environmental Protection Agency, Office of Radiation Programs, Waste Environmental Standards Program, Washington, D.C., February, 1978:*

---

* The writer is indebted to **Monthly Review,** June 1978, for its review of this report.

> It is generally conceded that risk estimates for many of the long-lived radionuclides would depend on numerous imprecise variables which would be little more than speculation after certain time periods. For example, Carbon-14, Plutonium-239 and Iodine-129 could present potential dose commitment for hundreds of thousands to millions of years. Such uncertainty could result in intense controversy over any calculations or judgements upon which environmental protection criteria or standards would be based. This circumstance places a considerable burden on government decisions regarding radioactive wastes. (page 1)

Are you and your government prepared to accept this "considerable burden" which should probably read "absolutely unsustainable burden"? Elsewhere the Report states:

> Other radionuclides which may be of special concern are I-129 and C-14. Iodine-129 is a volatile nuclide with a half-life of seventeen million years; thus its isolation from the biosphere over the long term will be difficult to assure. Carbon-14 has a half-life of 5,600 years and, if released to the biosphere, would represent exposure of the entire world's population due to rapid movement into the carbon cycle. (page 10)

It would not be inappropriate to change the words "difficult to ensure" to "totally impossible to assure". In another place the Report states:

> It should be noted... that there is no way to guarantee protection from pollutants such as radioactive materials which are assumed to have no threshold for effects, other than by prohibiting their production. (page 14)
>
> The degree of risk that the producing generation

There ***are*** *doubts about the safety of peaceful nuclear power reactors. The public interest in the matter is obvious and legitimate. The public has every excuse for being concerned.*

— Maurice Lamontagne, Chairman of the Select Committee on Science Policy of the Canadian Senate.

> passes on to the future represents an important legacy of radioactive wastes. This transference involves a moral judgement of responsibility, including the length of time for which responsibility extends into the future.* An ethical basis for decisions regarding risk transference is needed not only for philosophical reasons, but also for the practical purpose of implementing evaluation techniques and risk-cost analyses. Unfortunately, society has not established clear approaches for dealing with the imposition of such risks far into the future. (pages 17-18)

One could add that fortunately we do not need to assume that society will forever remain incapable of attaining a "clear" understanding of the situation and taking a stand against the imposition of such fundamental risks on future generations.

> Isolation of waste materials is an important consideration for reducing their risk-producing potential... In general, permanent isolation cannot be expected...

Or maybe, only for 5,600 years? See next quotation.

> Uncertainties dominate predicative ability. For more than a few thousand years, there is only uncertainty. (page 26)

> Deciding on a reasonable time for dependency on institutional controls is essentially equivalent to estimating the shortest likely period of institutional continuity. It is clear that there is no definitive scientific method for such a determination, but it is difficult to predict the course of events from current trends for more than several hundred years. (page 26)

In the history of the world are there any examples of such predictions for more than a few decades?

> Radiation effect on health at low doses and dose rates are presumed to have a linear dose-response

---

* In this connection it is reassuring to be told that "a study of the public attitudes and values associated with radioactive waste disposal based on polling techniques showed that long-term safety is widely held to be at least as important as safety over the short term." (page 16)

> relationship for public health protection purposes. This implies that any radiation dose, however small, conveys a proportionate chance of producing a genetic alteration or a somatic health effect such as cancer... Within this context, therefore, the appropriate goal is to avoid any increment of radiation in the general environment due to radioactive waste. (page 1)

When the Saskatchewan government and uranium companies undertake to embark in the growth and expansion of this unforgiving technology, you are inevitably contributing to the growth of wastes for which there is no disposal technology and for which there can be no disposal technology. As these wastes insinuate themselves into the global ecosystem, as they inevitably will, the lives of Saskatchewan people for thousands upon thousands of years will be as much at risk as the lives of the people of Japan, or Brazil or India or wherever they may live.

I repeat: there is no technological answer, there cannot be an answer and the tradeoff of short-term royalties for the evolutionary risk is not acceptable or justifiable from any moral perspective. Nuclear energy is forcing us at long last to recognize and admit the limits of technology and the need to relate our future social and economic developments to the scientific and moral knowledge arising from ecology.

In view of this, it is not surprising that church groups throughout the world have given leadership on this issue. Your statement at the press conference held to release the Bayda report that the churches do not have a monopoly on morality completely failed to address their moral concern. The churches do not have a conflict of interest on this issue as does your Government as part owner of the uranium exploration and mining industry.

The fission reactor produces both energy and radioactive waste: we want to use the energy now and leave the radioactive waste for our children and grandchildren to take care of.

— Dr. Hannes Alfven, Nobel physics laureate, 1974.

Conclusion

In your letter you presented six justifications for your decision to expand uranium mining in Saskatchewan. In this letter I have explored these points in some depth and shown them to be incredibly weak, virtually without foundation.

I am forced to the conclusion, therefore, that these **are not** the reasons why you decided to proceed. Then I am reminded of the many incidents before and during the Bayda Inquiry: the Cabinet support of uranium mining at the 1976 NDP Convention, the unusually short time you allowed for the Inquiry, the statements of Mineral Resources Minister Jack Messer during the Inquiry which implied that the Inquiry would not influence the decision in any case, the unconscionably short period between the time you received the report and announced your decision which left the public out of the process, etc. etc.

As you know, many people, including myself, believed you had made up your mind long ago. However, I had hoped for a more objective report from Bayda which would have forced you to at least consider the issues. The fact that Bayda dismissed serious consideration of virtually all issues put you in a position where you didn't have to deal with them.

This correspondence is, to my knowledge, the first place where you have really tried to present a justification, aside from your general endorsement of Bayda's recommendations. As you have failed to do so, I have had to search for the underlying reason behind your decision.

I now believe your decision flows from a basic commitment to the western industrial economic model. You are in no way committed to changing the direction of the economic forces controlled by the multinational corporations and their home governments. You are prepared to have the province become an active participant in this global economic system, only requiring "partnership" and "sharing" (those "lovely uranium dollars" as you say in your letter) with the multinational giants.

This concept of "socialism within one province" is based upon a fundamental fallacy. The ills of Saskatchewan are a result of the way in which the economic system operates. That economic system is global in nature and by participating in it as you are with uranium, you strengthen it, and by strengthening it you will multiply the problems of the province.

There are alternative ways to build a more self-reliant,

independent society with greater opportunities for participation, decentralization and democratic decision-making. This brings to mind the Robert Frost poem, so treasured by Woodrow Lloyd, which says

> Two roads diverged in a wood, and I —
> I took the one less travelled by,
> And that has made all the difference.

In this letter I have shown a number of ways in which the less travelled road is still open.

1. By declaring a moratorium on uranium mining for many reasons, but chiefly because of the unresolved public health and waste storage problems;

2. By developing a new approach to northern development which is based upon self-determination, respecting the aboriginal rights of native peoples;

3. By a study, and subsequent action, on the potential of socially useful employment based on alternative renewable energy policies;

4. By seriously investigating new research on the relationship between radiation and health and changing radiation standards accordingly as you have promised to do.

These are concrete ways of searching for alternatives to a global economic system which is responsible for the poverty and deprivation in Saskatchewan and in the world.

The public will support such initiatives. In fact, as they learn more about this terrifying technology, they will demand them. I add my voice to that growing demand.

Sincerely,

Bill Harding

Bill Harding

c.c. The public

Fission energy is safe only if a number of critical devices work as they should, if a number of people in key positions follow all their instructions, if there is no sabotage, no hijacking of transports, if no reactor fuel processing plants or waste repository anywhere in the world is situated in a region of riots or guerilla activity and no revolution or war — even a "conventional one" — takes place in these regions. The enormous quantity of extremely dangerous material must not get into the hands of ignorant people or desperados. No acts of God are permitted.

— Hannes Alfven, Nobel Prize physicist

"MR. SPEAKER... I REGRET TO SAY THAT SOME EMERGENCY LEGISLATION IS IN ORDER..."

Recommended Reading List

Nuclear Power, Walter C. Patterson, Penguin, 1976
The Empire of Oil, Harvey O'Connor, Monthly Review Press
The Seven Sisters, Anthony Sampson, Bantam Books, 1975
The Energy Crisis, Michael Tanzer, Monthly Review Press, 1974
Canada's Energy Crisis, James Laxer, James Laxer & Co., 1975
Middle East Oil and the Energy Crisis, Joe Stork, Monthly Review Press, 1975
Nuclear Power: The Fifth Horseman, Denis Hayes, Worldwatch Paper No. 6
Nuclear Energy: The Unforgiving Technology, Fred Knelman, Hurtig, 1976
Small is Beautiful, E.F. Schumaker, Abacus, 1973
The Poverty of Power, Barry Commoner, Knopf, 1976
Nuclear Power: The Bargain We Can't Afford, Richard Morgan, Environmental Action Foundation, 1977.
The Ethics of Uranium Development in Saskatchewan, Church in Society Committee, Sask. Conference, United Church of Canada, 1979.
Hazards of Nuclear Power, Allan Roberts and Zhores Medvedev, Spokesman, 1977.
Nuclear Power: The Unviable Option, John J. Berger, Ramparts Press, 1976.

Soft Energy Paths, Amory B. Lovins, Friends of the Earth, 1977

"Nuclear Energy in Canada", **Next Year Country**, April-May, 1977

"Saskatchewan Bonanza", **Next Year Country**, December, 1978

Periodicals

Nuclear Newsletter, Saskatoon Environmental Society, Box 1372, Saskatoon, Saskatchewan, S7K 3N9, $5.00 per year.

Briarpatch, 1409 - 10th Avenue, Regina, Saskatchewan, S4P 0E2, $8.00 per year

Alternatives, Trent University, Peterborough, Ontario, $4.00 per year

Bulletin of the Atomic Scientists, 1020 24e 58th Street, Chicago, Illinois 60637, $20.00 per year

Not Man Apart, Friends of the Earth, 124 Spear Street, San Francisco, California 94105, $10.00 per year

Critical Mass, P.O. Box 1538, Washington D.C. 20013, $15.50 per year

WIN, 503 Atlantic Avenue, Fifth floor, Brooklyn, New York 11217 $15.00 per year

Groundswell, Nuclear Information and Resource Service, 1536 Sixteenth Street NW, Washington, D.C., $12.00 per year

Seven Days, 206 Fifth Avenue, New York, N.Y. 10010 $9.00 for 15 issues

---

An excellent source for books and publications in Saskatchewan is

ONION BOOKS
650 Broadway Avenue
Saskatoon, Saskatchewan S7K 3L3

Interim Reply from Premier Blakeney's Deputy

Legislative Building
Regina, Saskatchewan
October 12, 1978.

Mr. Bill Harding
2230 Smith Street
Regina, Saskatchewan

Dear Bill:

I am writing on behalf of Premier Blakeney, and in his absence, to acknowledge and thank you for your open letter of September 20, 1978, concerning uranium development.

The Premier will be responding to you personally as soon as he is able to do so.

In the meantime, I thought I would pass along an excerpt from a letter which the Premier sent recently to someone else:

> "Any political party is made up of people who have different views on different subjects. I believe it was Dr. James Conant who said something to the effect that democracy represents a core of agreement among people surrounded by a rich profusion of diversity. The New Democratic Party is broadly based and will have the same attributes.
>
> Every citizen must reach his conclusions as to which

> Party best represents his views. If he fails to vote, he clearly gives support to those who do not represent his views. As has been so often said, for bad men to triumph it is necessary only for good men to do nothing. Even on an issue of nuclear power where no one of the Parties represents your view, it may be that one Party comes closer to your view than others. I would suggest that is the New Democratic Party."

This argument may not impress you, Bill, but it makes sense to me.

Yours sincerely,

Wes Bolstad

W.G. Bolstad,
Deputy Minister to the Premier.

Mr. Harding's Response

2230 Smith Street
Regina, Saskatchewan
November 28, 1978

W.G. Bolstad
Deputy Minister to the Premier
Legislative Building
Regina, Saskatchewan

Dear Wes:

Your letter of October 12 in reply to mine of September 20 arrived just prior to the election. It did nothing to convince me to vote. For the first time in my adult life I did not do so — and you will be aware of how deeply I would feel to permit that to happen.

Wes, we are clearly talking from two perspectives and I am not sure that the chasm can be crossed.

In my letter, I tried to say that, in my view, the uranium issue, because of its implications for human health and global existence, transcends the electoral process. In your reply, picking two paragraphs from "an excerpt from a letter which the Premier sent recently to someone else", you undertake to tell me that electoral politics is all-important, always. These are the two sides of the chasm we speak from.

I wish now to comment on those paragraphs of the Premier.

The first quoted paragraph talks about "the rich diffusion of diversity" within political parties. There is no great argument there except that it is an academic and idealistic description and overlooks the importance of the underlying guiding principles of the party, its leadership, communication within the party, and a host of other things. By itself it is rhetorical and virtually meaningless.

The second paragraph is quite another story.

First, it states: "Every citizen must reach his (her) conclusion as to which party best represents his (her) views." This is the essence of the entire matter, Wes. As stated above, when an issue as critical as uranium "development" is involved, and all parties are going to "develop" it, then there can be no party which "best represents his (her) views". The parties may have different formulae for "developing" the uranium but they will all do it. Their other differences — payment for health services, whether or not to nationalize, etc. are miniscule in the face of nuclear power and weaponry. This issue is so immense that all other issues, particularly the deceptive issue of "leadership" become completely superficial.

The second sentence states: "If he (she) fails to vote he (she) clearly gives support to those who do not represent his (her) views." The transparency of this statement is evident from what I have already said. Where there is **no** real difference on an underlying issue considered by the citizen to be fundamental to the existence of all human society, how can one political party be considered to "best represent" the citizen's views?

The third sentence states: "As has been so often said, for bad men (women) to triumph it is necessary only for good men (women) to do nothing." Come now, Wes! The context in which this sentence appears implies that the only thing a person can do is vote. You know there are many other ways outside the electoral process for people to participate if a society is truly democratic. For my part meaningful participation is a day-to-day responsibility and always has been. Voting is something like a period to a sentence. Our constant resistance to uranium development is no less a contribution to the future of society than for example, Al Blakeney's running for Premier. In fact, Wes, I will be happy to accept history's judgement on that comparison!

The final two sentences of the second paragraph require no further comment.

In summary, Wes, the second paragraph is insultingly shallow.

It preaches. It attempts to lay a guilt trip on people. It creates "good" and "bad" groups. It is self-righteous, parochial and divisive.

However, the real weakness of the letter is that it didn't deal with my reasons for not voting. So let me give you just two of those reasons again:

1. Mining uranium anywhere will add to the risk of nuclear war.
2. Mining uranium anywhere will produce radioactive wastes which will add unacceptable levels of cancer, leukemia and cellular damage to our and future generations.

As I said in my letter, Bayda did the Government and the public a real disservice by refusing to look at these and other questions seriously. We are presently documenting his failures and will be reporting soon. It will not be too late for the Government to review its position.

Finally, Wes, I think the greatest disservice to the public was that this was not permitted to be an issue in the campaign. I say "not permitted" advisedly because an ad presenting the non-nuclear side of the issue was refused publication by the **Commonwealth** (see my letter to the Editor and copy of the ad in the November 15 issue). This ad was sponsored by 37 organizations including the New Democratic Women. In view of this, I think it was a shameful commentary on the present state of public affairs when the Premier was able to say, the day following the election, "Uranium was not an issue in the campaign". All the goodies promised by all parties were to be gained from natural resources, primarily from uranium, and yet "uranium was not an issue in the campaign". This may be an effective way to get elected but, to me, it turns the election into a combination con-game and lottery.

Parliamentary democracy has come to a sorry pass, Wes, when we fail to take people into our complete confidence. This has not been done in the case of uranium development. In fact it is seldom done in our increasingly centralized, bureaucratized and monopoly-controlled society. The recent exposure of the PCB spill in Regina, held secret from the public for two years, is another example of how politicians and bureaucrats fundmentally disrespect and distrust the public. When this trust disappears then electoral politics becomes irrelevant. This irrelevance was the outstanding characteristic of the recent election because no party had the strength to raise the underlying issue of uranium "development". In the short, four year time span of politicians, I can understand why the NDP wanted the issue suppressed but that doesn't make me feel any better.

In future, Wes, I would hope for a more substantive reply.

Sincerely,

Bill Harding

A society that blindly accepts the decisions of experts is a sick society on its way to death.

— René Dubos

The Newspaper Advertisement

Say "NO" to uranium mining

Nuclear power is not needed

The world doesn't need nuclear power. New oil and gas reserves are discovered regularly, which, with intelligent conservation programs, will last well over 100 years. Solar, wind and biomass energy are all safe, clean and sufficient for our real needs in an energy efficient future. In addition, these everlasting sources of energy have been shown to be capable of creating more employment than would development of more capital-intensive energy sources.

Before the government report on uranium was released, Premier Blakeney stated that Saskatchewan could get along just fine without future expansion of uranium mining.

Uranium sales will not bring us riches

Uranium companies everywhere are cashing in on the present high price of uranium set by their cartel. But the nuclear industry is already slowing down because of excessive inflation and fundamental safety and waste disposal problems.

Over the last three years there have been twice as many cancellations of orders as there have been new orders for nuclear power plants in the U.S.

No major royalties will come into the provincial government treasury until the industry has recovered all its costs. Yet our taxes will be used to build all the support services (roads, etc.) to enable this foreign-owned industry to expand in Saskatchewan. When the uranium market collapses, the people of Saskatchewan (like the people of Sudbury) will be abandoned by corporations who have profited, without taking any risks, while being subsidized by our tax dollar.

Uranium mining is not northern development

If northern Saskatchewan's uranium resources are extracted, the bulk of the benefits will go to the mining and construction companies. Other resource extraction projects in northern Canada have shown that Northerners gain little from such projects and that native people in particular suffer a negative social impact and a further loss of independence.

With the construction of new mines will come high wages for a few, along with disruption of the existing economy, increased prices, more alcoholism and prostitution. As the profits flow out, the North will further "underdevelop".

Uranium mining increases the risk of nuclear war

Uranium is the basic material for nuclear weapons, and the expansion of the nuclear industry is increasing the chance of a nuclear war. Ninety percent of Canadian uranium now goes into the world market where it can end up supplying unstable and authoritarian governments with nuclear weapons material. (Amok, the company given the go-ahead to mine uranium, is from France, which has **not** signed the nuclear non-proliferation treaty, and has holdings in such countries as South Africa.)

The limitations and weaknesses of the present safeguards "are so serious that the safeguards may provide only an illusion of protection." — Report of the Australian Royal Commission on Uranium Mining, 1976.

No safe disposal method exists for radioactive wastes

Every year there is more radioactive Strontium, Iodine and Cesium produced inside a single nuclear power reactor than has been produced in the fallout of all atomic bomb tests to date. These materials are extremely toxic for thousands of generations and must not be allowed to enter the environment of living things.

The U.K. Royal Commission Report on Nuclear Energy (1976) states, "We are agreed that it would be irresponsible and morally wrong to commit future generations to the consequences of fission power on a massive scale unless it has been demonstrated beyond a reasonable doubt that at least one method exists for the safe isolation of these wastes for the indefinite future."

Uranium mining is hazardous to our health

Uranium miners at Elliot Lake got lung cancer at a rate double that of the general public. — Ontario Royal Commission on the Health and

Safety of Workers in Mines, 1976.

Exposure to levels of radiation below "permissable" levels has been shown to cause a variety of health problems including cancer and leukemia. Recent studies in the U.S. indicate higher rates of cancer and leukemia downwind from nuclear power stations.

Didn't the Bayda Inquiry give these issues serious consideration?

No, it didn't. A good example is the concern that Saskatchewan's uranium could end up in atomic bombs. The Inquiry agreed that this was a serious concern, but like Pontius Pilate, they "washed their hands of it", by saying Saskatchewan wasn't that influential anyway. With that attitude, why did they bother to conduct an inquiry?

The effect of exposure to low level radiation is another example of an overlooked concern. New research, published after the inquiry closed its hearings shows that small amounts of radiation over a long period of time are much more hazardous than was previously believed. The inquiry was in such a rush that it did not deal with this important new scientific information.

The biggest failure of the Government in setting up the Bayda Inquiry was in neglecting to learn what the people of this province have to say about uranium mining. The fact is that a Gallup poll released in December 197[illegible] showed that opposition to nuclear power is large and growing larger. The future of the world is too important to be left to a tiny group in industry and government.

Time must be allowed for everyone in Saskatchewan to become fully informed and to give their opinion on the nuclear issue.

The opening of new uranium mines is against the public interest and world interest!

These organizations support this statement

- Saskatoon Environmental Society • Women Against Nuclear Development • Market Square Food Co-op
- Canadian University Services Overseas (CUSO), Saskatoon Local Committee • Inter-Church Energy Committee
- Saskatchewan Association of Social Workers • Regina Group for a Non-Nuclear Society (RGNNS)
- Voice of Women • Saskatchewan Joint Board, Retail Wholesale and Department Store Union (RWDSU)
- Warman and Area Concerned Citizens • University Coalition Against Nuclear Development (Saskatoon)
- Church in Society, Saskatchewan Conference, United Church of Canada • University of Regina Student Union
- Grain Services Union (CLC) • Association of Metis and Non-Status Indians of Saskatchewan (AMNSIS)
- Saskatchewan Coalition Against Nuclear Development • Saskatoon Community Clinic
- Oxfam (Saskatoon Local Committee) • Regina Friends Meeting (Quakers) • Magpie Symbiotic Co-operative
- Saskatchewan Cross-Cultural Centre • Saskatchewan Association on Human Rights
- Student Christian Movement • Briarpatch magazine • New Democratic Women
- Church in Society Committee, Saskatoon Presbytery, United Church of Canada • Farm-Maid Products Ltd.
- Regina Women's Community Centre • Saskatoon Community Clinic Medical Group
- Board of Directors, Saskatchewan Council for International Co-operation (SCIC)
- Healthy Horizons • Canadian University Services Overseas (CUSO), Regina Local Committee
- Regina Committee for World Development (RCWD) • Women's Action Collective on Health (WACH)
- Saskatchewan Union of Nurses (Local 101) • Oxfam (Regina Local Committee)
- Tip of the Iceberg, Training/Organizing Collective

Say "YES" to a safe and sound energy future!

- ☐ I support the above statement.
- ☐ I would like to be kept informed about this issue.
- ☐ I enclose $______ to assist efforts to achieve a non-nuclear society.
- ☐ I would like to volunteer my time to help with this issue.
- ☐ I enclose a letter with other comments.

Name ______________________________

Address ______________________________

City or town ____________________ **Postal code** __________

Mail to: Saskatoon Environmental Society, Box 1372, Saskatoon, Sask. S7K 3N9
or: Regina Group for a Non-Nuclear Society, 2138 McIntyre St., Regina, Sask. S4P 2R7

The great fault of all ethics hitherto has been that they believed themselves to deal only with the relation of man to man.

— Albert Schweitzer

Letters to 'the Commonwealth'

— Reprinted from the *Commonwealth*, Saskatchewan NDP newspaper, November 15, 1978

Commonwealth accused of censorship

On September 14 our representative phoned you about reserving space in The Commonwealth for a full-page ad on the uranium issue sponsored by 37 organizations in the province. You agreed to reserve a page in the October 4 issue.

The camera-ready copy was delivered to you on September 20 and you confirmed that it would be carried in the October 4 issue. You were given a cheque for the cost of placing the ad.

When we received the October 4 issue, the ad was not in it. Our representative called you immediately requesting an explanation. You replied that you had meant to tell us the ad had been cut but hadn't got around to it. You said that "after discussing it with the editorial board we decided not to run it", that it was "too dicey", there was "no way it was going to go during an election". It was "not a question of wanting to whitewash the government," but the ad was "inopportune during an election campaign."

Your agreement, on September 14 and 20, to run the ad, followed by your change of mind and neglecting to tell us about it in our view constitutes an act of censorship. It was discriminatory to the 37 organizations who wished to make their views known, and unfair to the general public who have the right to know responsible opinions on the subject. We paid in advance so they could know.

Premier Blakeney is quoted as saying: "Uranium was a non-issue in the campaign" (Regina Leader-Post, October 20, 1978, page 9). It was clearly the intent of the NDP officials to ensure that this critical question would be a non-issue. We have come to expect this from the establishment media, but that The Commonwealth was used to help achieve this end is unfortunate to say the least.

I trust that you will publish this letter and thus not double your sin of censorship. I also request you to now publish the ad. You already have our cheque for $400-plus contributed by the 37 organizations involved. The people who read The Commonwealth have the right to know the reasons for the growing opposition to uranium exploitation.

W.M. Harding
for the Regina Group
for a Non-Nuclear Society

— Reprinted from the *Commonwealth*,
December 13, 1978

Censuring the censors

If the facts in W.M. Harding's letter of November 15, 1978 are correct, then The Commonwealth was guilty of censorship in its handling of the "We Say No To Uranium Mining" ad. I am disappointed.

The organizations sponsoring the ad clearly wanted to run it during the provincial election and The Commonwealth should have complied. During elections, issues are more fully debated than at other times and decisions are made. The left in Canada has lost much in the past from censorship decisions made by the owners of the media. It is sad that The Commonwealth would consciously restrict debate during an election in the same way.

The ad itself is very informative on the issue. It did not attack the Saskatchewan NDP government nor advocate voting for its opponents.

Can the editors of The Commonwealth give us their side of this episode. What happened and why?

John Smart
Ottawa

Editor's note: What happened was fairly represented by Mr. Harding in his letter.

Why it happened can best be put down to the practice of **real politik** — the difference between the way we would have things be and the way they really are.

*The **malaise** I have come more and more to believe, lies in the industrial basis on which our civilization rests. Economic growth and technical achievement, the greatest triumphs of our epoch of history, have shown themselves to be inadequate sources for collective contentment and hope. Material advance, the most profoundly distinguishing attribute of industrial capitalism and socialism alike, has proved unable to satisfy the human spirit. Not only the quest for profit but the cult of efficiency have shown themselves ultimately corrosive for human well-being. A society dominated by the machine process, dependent on factory and office routine, celebrating itself in the act of individual consumption, is finally insufficient to retain our loyalty.*

— Robert L. Heilbroner, Second Thoughts on *The Human Prospect.*

RGNNS

The Regina Group for a Non-Nuclear Society (RGNNS) is a voluntary organization of citizens in southern Saskatchewan who share a concern about the present direction of society. We consider the most threatening current example of this direction to be the stubborn efforts of industry and government to establish a nuclear power industry throughout the world. We have come together to combat this trend and to encourage the growth of an energy policy based on renewable resources and the wise use of all resources.

RGNNS believes, however, that the urgent need **now** is to inform the public about the serious hazards of a uranium mining/nuclear power type of society. The Government and industry are engaged in a public relations effort to sell such a society. They are not telling the whole story.

RGNNS accepts its share of the responsibility to inform the public. This publication is the first of a series. Others will provide a detailed critique of the Bayda Inquiry Report, on which the Provincial Government made its decision to expand uranium mining in northern Saskatchewan.

We need the help of all readers to spread this information. Donations to help pay the cost of publishing are needed. Most important is to buy and distribute copies of this book to friends and neighbours, and to write letters to newspapers, MLA's and MP's. Please keep in touch with us.

Regina Group for a Non-Nuclear Society (RGNNS)
2138 McIntyre Street
Regina, Saskatchewn S4P 2R7

Former CCF/NDP colleagues Allan Blakeney, Premier of Saskatchewan, and Bill Harding, retired United Nations senior official, disagree sharply on the wisdom of continuing the mining of uranium in Saskatchewan.

In the enclosed correspondence, Premier Blakeney presents, for the first time since announcing that the Cluff Lake uranium mine will proceed, his justifications for the expansion of uranium mining.

Mr. Harding looks at each justification in depth, drawing on his wide experience in development in Saskatchewan and around the world. He challenges the Premier's arguments and exposes their superficial nature and their fundamental lack of foresight.

Calling for an alternative energy policy based on renewable energy, Mr. Harding claims this will lead to a non-colonialist technology which will encourage greater community self-reliance. It will create more jobs — jobs which are more socially useful. It will create an ecologically inoffensive type of society.

High priority is placed on settlement of aboriginal land claims to ensure self-determination for native people.

Mr. Harding maintains that government and industry are not following people-oriented development policies. He believes present policies lock Saskatchewan more firmly into a global economic system which creates and deepens underdevelopment and poverty. He points out the fallacy of the Premier's "golden egg" approach and identifies the unreality of attempting to create "socialism within one province".

Essential Reading for Concerned Citizens